AF800681

Biosystems & Biorobotics 35

Series Editor

Eugenio Guglielmelli, *Laboratory of Biomedical Robotics, Campus Bio-Medico University of Rome, Rome, Italy*

Editorial Board Members

Dino Accoto, *Università Campus Bio-Medico di Roma, Rome, Italy*
Sunil K. Agrawal, *Department of Mechanical Engineering, Columbia University, New York, USA*
Fabio Babiloni, *BrainSigns, University of Rome Sapienza, Rome, Italy*
Jose Carmena, *University of California Berkeley, Berkeley, CA, USA*
Maria Chiara Carrozza, *Biorobotics Institute, Scuola Superiore Sant'Anna, Pisa, Italy*
Paolo Dario, *BioRobotics Institute, Sant'Anna School of Advanced Studies, Pisa, Italy*
Arturo Forner-Cordero, *University of Sao Paulo, Sao Paulo, Brazil*
Masakatsu G. Fujie, *Waseda University, Tokyo, Japan*
Nicolas Garcia, *Ingeniería de Sistemas y Automática, Universidad Miguel Hernández, Elche (Alicante), Spain*
Neville Hogan, *Mechanical Engineering, Brain and Cognitive Sciences, Massachusetts Institute of Technology, Cambridge, MA, USA*
Hermano Igo Krebs, *Newman Laboratory for Biomechanics and Human Rehabilitation, Massachusetts Institute of Technology, Cambridge, MA, USA*
Dirk Lefeber, *Dep. of Mechanical Engineering, Univ. of Bruxelles, Bruxelles, Belgium*
Rui Loureiro, *University College London, London, UK*
Marko Munih, *Faculty of Electrical Engineering, University of Ljubljana, Ljubljana, Slovenia*
Paolo M. Rossini, *University Cattolica del Sacro Cuore, Rome, Italy*
Atsuo Takanishi, *Department of Modern Mechanical Engineering, Waseda University, Tokyo, Japan*
Russell H. Taylor, *Department of Computer Science, Johns Hopkins University, Baltimore, MD, USA*
David A Weitz, *Department of Physics & SEAS, Harvard University, Cambridge, MA, USA*
Loredana Zollo, *Università Campus Bio-Medico di Roma, Roma (RM), Italy*

The BIOSYSTEMS & BIOROBOTICS (BioSysRob) series publishes the latest research developments in three main areas: 1) understanding biological systems from a bioengineering point of view, i.e. the study of biosystems by exploiting engineering methods and tools to unveil their functioning principles and unrivalled performance; 2) design and development of biologically inspired machines and systems to be used for different purposes and in a variety of application contexts. In particular, the series welcomes contributions on novel design approaches, methods and tools as well as case studies on specific bio-inspired systems; 3) design and developments of nano-, micro-, macro- devices and systems for biomedical applications, i.e. technologies that can improve modern healthcare and welfare by enabling novel solutions for prevention, diagnosis, surgery, prosthetics, rehabilitation and independent living. On one side, the series focuses on recent methods and technologies which allow multi-scale, multi-physics, high-resolution analysis and modeling of biological systems. A special emphasis on this side is given to the use of mechatronic and robotic systems as a tool for basic research in biology. On the other side, the series authoritatively reports on current theoretical and experimental challenges and developments related to the "biomechatronic" design of novel biorobotic machines. A special emphasis on this side is given to human-machine interaction and interfacing, and also to the ethical and social implications of this emerging research area, as key challenges for the acceptability and sustainability of biorobotics technology. The main target of the series are engineers interested in biology and medicine, and specifically bioengineers and bioroboticists. Volume published in the series comprise monographs, edited volumes, lecture notes, as well as selected conference proceedings and PhD theses. The series also publishes books purposely devoted to support education in bioengineering, biomedical engineering, biomechatronics and biorobotics at graduate and post-graduate levels.

Indexed by SCOPUS, WTI Frankfurt eG, SCImago

Monica Carfagni · Simona Celi · Michaela Servi ·
Marta Mencarelli
Editors

3D Printing and Bioprinting in Medicine and Surgery

Proceedings of the 4th National Congress of the Italian Digital Biomanufacturing Network, September 26–27, 2024, Florence, Italy

Springer

Editors
Monica Carfagni
Department of Industrial Engineering
University of Florence
Florence, Italy

Michaela Servi
Department of Industrial Engineering
University of Florence
Florence, Italy

Simona Celi
Gabriele Monasterio Tuscany Foundation
Massa, Italy

Marta Mencarelli
Department of Industrial Engineering
University of Florence
Florence, Italy

ISSN 2195-3562 ISSN 2195-3570 (electronic)
Biosystems & Biorobotics
ISBN 978-3-031-94625-7 ISBN 978-3-031-94626-4 (eBook)
https://doi.org/10.1007/978-3-031-94626-4

© The Editor(s) (if applicable) and The Author(s), under exclusive license
to Springer Nature Switzerland AG 2025

This work is subject to copyright. All rights are solely and exclusively licensed by the Publisher, whether the whole or part of the material is concerned, specifically the rights of translation, reprinting, reuse of illustrations, recitation, broadcasting, reproduction on microfilms or in any other physical way, and transmission or information storage and retrieval, electronic adaptation, computer software, or by similar or dissimilar methodology now known or hereafter developed.
The use of general descriptive names, registered names, trademarks, service marks, etc. in this publication does not imply, even in the absence of a specific statement, that such names are exempt from the relevant protective laws and regulations and therefore free for general use.
The publisher, the authors and the editors are safe to assume that the advice and information in this book are believed to be true and accurate at the date of publication. Neither the publisher nor the authors or the editors give a warranty, expressed or implied, with respect to the material contained herein or for any errors or omissions that may have been made. The publisher remains neutral with regard to jurisdictional claims in published maps and institutional affiliations.

This Springer imprint is published by the registered company Springer Nature Switzerland AG
The registered company address is: Gewerbestrasse 11, 6330 Cham, Switzerland

If disposing of this product, please recycle the paper.

Preface and Acknowledgements

This book gathers the proceedings of the 4th National Congress on 3D Printing and Bioprinting in Medicine and Surgery (IDBN 2024), held at the Meyer Health Campus in Florence, Italy, on September 26–27, 2024. The event is organized by the Department of Industrial Engineering of Florence and the Gabriele Monasterio Foundation of Massa.

Building on the success of previous editions, IDBN 2024 provides a highly relevant platform for the scientific community to exchange knowledge and insights into the rapidly evolving fields of 3D printing and bioprinting. These technologies play a pivotal role in contemporary biomedical research and clinical applications, fostering interdisciplinary collaboration and technological advancement.

Experts in the field of biosystems and biorobotics, coordinated by the General Chairs and supported by a high-level Technical and Scientific Committee, were involved in the review process, which was conducted through a double-blind procedure and led to the selection of 11 papers. Contributions focus on innovative methodologies, cutting-edge technologies, and the latest research findings in these domains, promoting integration between different medical disciplines and state-of-the-art engineering approaches.

The editors would like to personally thank everyone participating in the review process for their dedication and skill at the service of the Conference. Special appreciation goes to the members of the Technical and Scientific Committees who made it possible for the Conference to take place.

The Conference prestige was enhanced by the participation of the keynote speakers: Prof. Bernardo Innocenti, Dott. Flavio Facchini, Dott. Federico Mussa. General Chairs extend their thanks to all authors, speakers, and those persons whose labour, financial support and encouragement made the IDBN 2024 event possible.

Monica Carfagni
Simona Celi
Michaela Servi
Marta Mencarelli

Organization Committee

General Chairs

Monica Carfagni of the University of Florence
Simona Celi of the Monasterio Foundation

Organizing Committee

Monica Carfagni	University of Florence
Michaela Servi	University of Florence
Marta Mencarelli	University of Florence
Simona Celi	Monasterio Foundation
Katia Capellini	Monasterio Foundation
Benigno Marco Fanni	Monasterio Foundation
Emanuele Gasparotti	Monasterio Foundation
Alberto Leardini	Rizzoli Orthopaedic Institute-IRCCS

Technical and Scientific Committee

Monica Carfagni	Department of Information Engineering, University of Florence
Simona Celi	Monasterio Foundation
Michaela Servi	Department of Information Engineering, University of Florence
Marta Mencarelli	Department of Information Engineering, University of Florence
Francesco Buonamici	Department of Information Engineering, University of Florence
Emanuele Gasparotti	Monasterio Foundation
Katia Capellini	Monasterio Foundation
Benigno Marco Fanni	Monasterio Foundation
Federico Mussa	AOU Meyer-IRCCS
Mirko Scagnet	AOU Meyer-IRCCS

Directive Board

Paolo Domenico Parchi	Orthopaedic and Traumatology Clinic AOU PISANA
Alberto Leardini	Rizzoli Orthopaedic Institute-IRCCS
Maria Livia Rizzo	Stefanelli & Stefanelli Law Firm
Ferdinando Auricchio	University of Pavia
Nicola Bizzotto	Bressanone Hospital

Contents

From ADAM to ADAM⚥: Design and Development of a 3D Printed Modular Respiratory Phantom for Radiotherapy Treatments 1
Andrea Profili, Michaela Servi, Yary Volpe, and Stefania Pallotta

GelMA-Xanthan Gum Hydrogel Bioink Preserving hiPSCs Viability and Proliferation 9
Virginia Deidda, Isabel Ventisentte, Marianna Langione, José Manuel Pioner, Caterina Credi, and Federico Carpi

Development and Implementation of a Pediatric Simulator for Training in Intraosseous Access Procedure 17
Antonio Marzola, Marta Mencarelli, Lorenzo Torzini, Marco De Luca, Monica Carfagni, and Yary Volpe

Prototyping and Testing of a Novel Simulator of Congenital Heart Diseases for Hands-on Surgical Training 25
Michele Bertolini, Alessandro Maria Marchini, Neel Murthy, Francesco De Gaetano, Luke Smith, Mirjana Cvetkovic, Giorgio Colombo, and Claudio Capelli

Direct Ink Printing of Liquid Crystalline Elastomers Towards Photoresponsive Artificial Muscles 33
Greta Feregotto, Giovanni Simonetti, Ruggero Rossi, Daniele Martella, Caterina Credi, Camilla Parmeggiani, and Federico Carpi

Surface Characteristics Improvement of 3D Printed 316 L Steel for Biomedical Applications 45
Eleonora Bolli, Saulius Kaciulis, Alex Lanzutti, Alessio Mezzi, Roberto Montanari, Alessandra Palombi, Francesco Sordetti, Emanuele Vaglio, Alessandra Varone, Claudio Verona, and Ali Zahid

Microstructural Stability of Co28Cr6Mo Biocompatible Alloy Produced by Additive Manufacturing 53
Alessandra Fava, Saulius Kaciulis, Alessio Mezzi, Roberto Montanari, Alessandra Palombi, and Alessandra Varone

Advanced Biomechanical Modelling Techniques and Materials' Analysis in Dental Structures 61
Giuliana Baiamonte, Fulvia Monaco, Giuseppe Laudani, Gianfranco Di Martino, Riccardo Asero, and Michele Calì

Development and Testing of a Patient-Specific Cardiovascular Simulator for Interventional Valve-in-Valve Procedure Training 69
Emanuele Gasparotti, Benigno Marco Fanni, Roberta Zola, Chiara Corno, Massimiliano Mariani, Sergio Berti, and Simona Celi

Design and Fabrication of 3D Printed Deformable Model of Patient-Specific Left Atrium for PIV Investigation 81
Emanuele Gasparotti, Benigno Marco Fanni, Emanuele Vignali, Mattia Dell'Amico, Rachele Marangoni, Francesca Danielli, Francesca Berti, Guido Parodi, Sergio Berti, Giancarlo Pennati, Lorenza Petrini, and Simona Celi

CAD-Driven Design for 3D-Printed Surgical Guides and Cranial Implants 95
Francesco Buonamici, Monica Carfagni, Federico Mussa, Agushi Rina, and Antonio Marzola

Author Index 103

From ADAM to ADAM ⚥: Design and Development of a 3D Printed Modular Respiratory Phantom for Radiotherapy Treatments

Andrea Profili[1](✉), Michaela Servi[1], Yary Volpe[1], and Stefania Pallotta[2,3]

[1] Department of Industrial Engineering, University of Florence, Florence, Italy
andrea.profili@unifi.it

[2] Department of Biomedical, Experimental and Clinical Sciences "Mario Serio", University of Florence, Florence, Italy

[3] AOU Careggi, Florence, Italy

Abstract. In radiotherapy treatments of mobile lesions, several strategies are used to treat the tumour while sparing the surrounding tissues [1]. The ADAM phantom (Anthropomorphic Dynamic breAthing Model) was developed to test dose delivery in lung cancer treatment. The aim of this study is to create ADAM ⚥(Anthropomorphic Dynamic breAthing Model - ⚥male&female), a new modular phantom based on ADAM, to test free-breathing and deep inspiration breath hold (DIBH) breast radiotherapy treatments [2, 3]. The 3D model of ADAM was reconstructed using computed tomography (CT) scans to ensure a high level of anatomical accuracy. This digital model of ADAM served as the starting point for the design and development of ADAM ⚥. 3D printable materials with tissue mimicking properties were selected to replicate tissues and organs [4–9]. The torso and chest were printed in PLA and the ribs and spine were printed in PLA StoneFil™ [10]. Tests have shown that ADAM ⚥closely mimics human tissue in terms of Hounsfield Units. Its modular structure allows rapid transition between male and female configurations, making it suitable for realistic simulation of many clinical situations. ADAM ⚥demonstrates adequate performance for future testing gating and tracking devices in the treatment of moving breast and lung lesions.

Keywords: Anthropomorphic Breathing Phantom · 3D Printing · 3D Modelling · Tissue Mimicking

1 Introduction

Radiotherapy uses ionising radiation to destroy tumour cells [1]. However, accurately targeting the dose to the tumour while reducing damage to surrounding healthy tissue is a challenge, especially for mobile lesions such as lung lesions where respiratory movement causes changes in tumour position. To address this complexity, synchronisation techniques and tracking devices have been developed to follow tumour movement and adjust radiation delivery. The ADAM phantom is an innovative device for testing and

© The Author(s), under exclusive license to Springer Nature Switzerland AG 2025
M. Carfagni et al. (Eds.): IDBN 2024, BioSysRob 35, pp. 1–8, 2025.
https://doi.org/10.1007/978-3-031-94626-4_1

calibrating these systems by simulating human respiratory motion [2, 3]. ADAM has improved lung cancer radiotherapy techniques, but the need for more versatile models led to the development of ADAM ⚥. This new ADAM-based phantom is suitable for testing free-breathing and deep inspiration breath hold (DIBH) breast radiotherapy treatments and faithfully replicates the anatomical conditions and respiratory movements of male and female patients.

2 Materials and Methods

The development of the ADAM ⚥phantom consisted of three stages. First, it was necessary to understand the hardware structure able to produce the breathing movement of ADAM. This was then used to analyse the type of movement for ADAM ⚥. Then the anatomical parts of the phantom were modelled in order to use the components already present in ADAM. Finally, an analysis of materials was carried out to identify those with tissue mimicking properties that would be suitable for the construction of the phantom.

2.1 Simulating Respiratory Movement

The simulation of respiratory movement in ADAM is achieved using three stepper motors (2), (6) and (10) as shown in Fig. 1b. Motors (6) and (2) are responsible for the movement of the lungs (1), which are connected to the main structure (4) by carbon rods (8). Motor (10) controls the movement of the anterior chest wall (hereinafter referred to as hatch). As can be seen in Fig. 1a, the ADAM hatch pivots on a plastic joint that acts as a cylindrical joint, which allows the hatch to rotate around the neck area thanks to the actuator (10) shown in Fig. 1b.

The movement that the ADAM hatch makes is therefore a rotary movement around the joint on the neck. From clinical requirements, unlike ADAM, the movement of the female phantom chest walls must be a translatory movement perpendicular to the coronal plane. This different type of movement stems from the necessity to compare measured dose distributions with planned ones. For breast treatment, it's crucial to measure dose on an axial plane, which is why a film housing on such a plane has been designed. To preserve the plane's orientation, it's essential for the entire chest wall to move vertically without rotation around a joint (unlike ADAM), which would otherwise alter the plane's original orientation. To achieve this type of movement, it was decided to use motor (2) to obtain translational movement with respect to the vertical axis. Therefore, a new support was designed to be placed on the carbon rods to accommodate the female hatch.

To accommodate the need for two distinct types of chest walls -male and female- it was decided to design the phantom with maximum modularity. This allows for swift transitions between ADAM ⚥male and female configurations as required.

2.2 3D Reconstruction and Modelling

A CT study of the physical ADAM phantom (shown in Fig. 1) was used to create the 3D model. CT data was processed using 3D modelling software such as Mimics (Materialise) and Geomagic Design X to create a detailed digital model of the anatomy involved. The

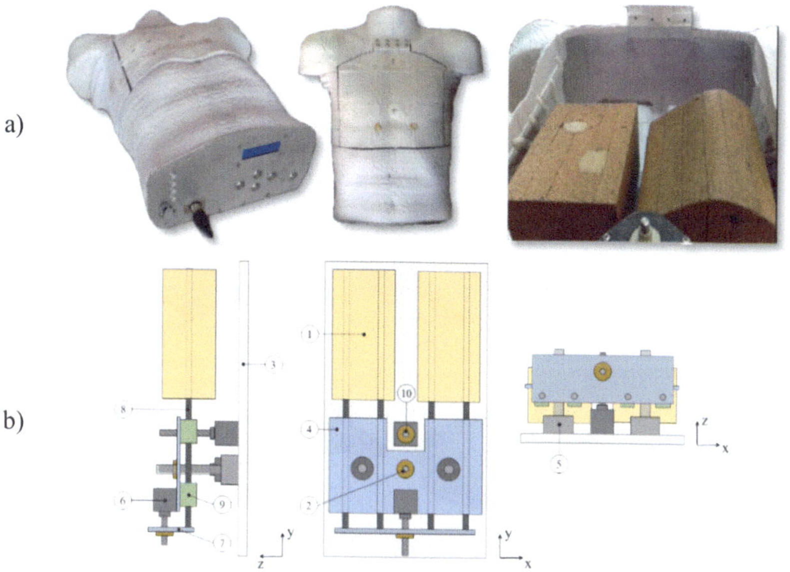

Fig. 1. a) Physical structure of ADAM phantom; b) Sketch of the electro-mechanical device located in the caudal region of the ADAM phantom and responsible for the movement of the lungs (1). One linear actuator is (2) fixed to the lower wall (3) of the phantom and move a U-shaped bar (4) up and down, along two guides (5). The second actuator (6), fixed to the U-bar, moves the iron plate (7), in the cranial-caudal direction. Four carbon tubes (8) are attached to the iron plate and lungs, they slide in cylindrical guides (9). The third actuator (10) moves the anterior chest wall up and down.

digital model was then used as the starting point for modelling the new phantom. Once the digital model was obtained, a detailed review was performed to ensure that all critical anatomical structures were accurately represented.

Firstly, as shown in Fig. 2a–b, the pre-existing cut for the male respiratory tract was removed to rejoin the thorax to create a single piece, and then the phantom surface was smoothed. The next step was to change the geometry of the phantom. In fact, ADAM mimics a position where the arms are stretched out at the sides and then down. For breast cancer radiotherapy, on the other hand, each patient is made to lie on their back with their arms up. For the ADAM ⚥phantom, the anatomy was then modified by importing the model of an appropriate female CT and merging the two models with a Boolean function between the meshes, as shown in Fig. 2c. In addition, to obtain a more realistic result, it was decided to model the cardiac area in two different ways. For the male phantom, given the presence of the lungs inside the torso, it was decided not to change the existing representation, i.e. a triangular area (blue in Fig. 2d); for the female phantom, on the other hand, a more realistic heart was modelled (green in Fig. 2d), to test dose distribution in a more realistic scenario.

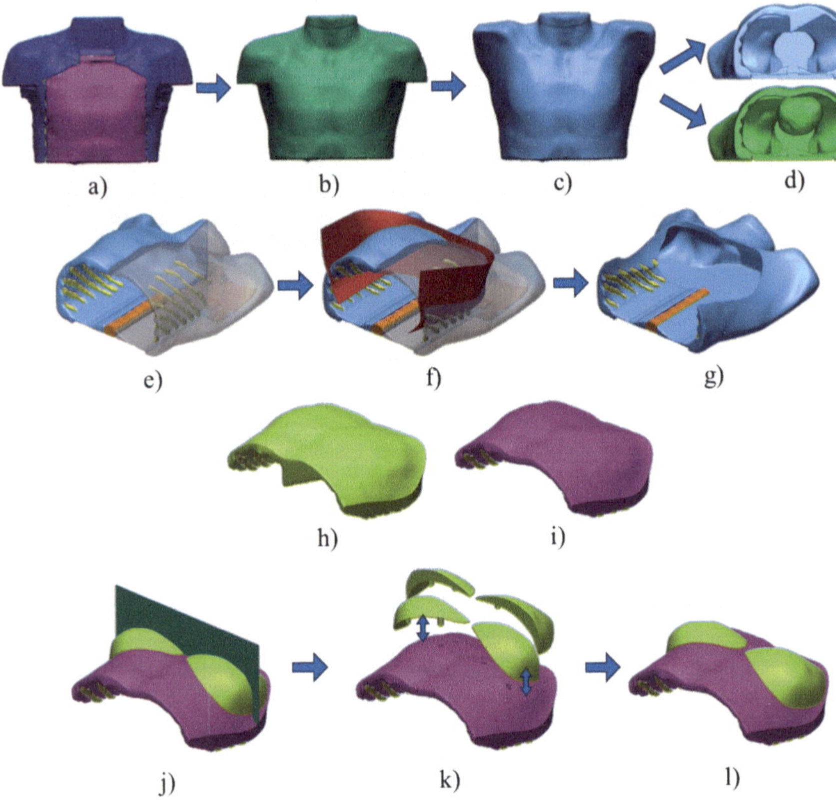

Fig. 2. From 3D ADAM reconstruction to ADAM ⚥design.

As shown in Fig. 2c–d, the torso model is the same for both phantom models, except for the area dedicated to the representation of the heart structure. The bone tissues, namely the spine and ribs, were then modelled. For the previous ADAM phantom, the ribs were modelled with a calcium carbonate compound, while the spine was modelled with a plastic tube and calcium carbonate. New ribs were then modelled (Fig. 2e), based on the existing ADAM, for both male and female phantoms. The spine was also modelled by adding more detail to simulate the real structure, as shown in Fig. 2e.

As introduced in Sect. 2.1, the simulation of the breathing movement in the phantom is done by moving the hatch. In order to ensure better modularity, it was decided to make a cut with the same surface (Fig. 2f) for both chests. In this way it is possible to obtain two different hatches (male one Fig. 3h, female one Fig. 3i) that will simulate the movement, while the rest of the chest remains motionless and same for both phantoms.

Finally, at the specific request of the clinical staff, a split plane was created through the heart and chest into which a detector film could be inserted for dose testing, as shown in Fig. 2j. It was then decided to make the breasts modular as well, creating cylindrical guides for insertion from above, as shown in Fig. 2k-l.

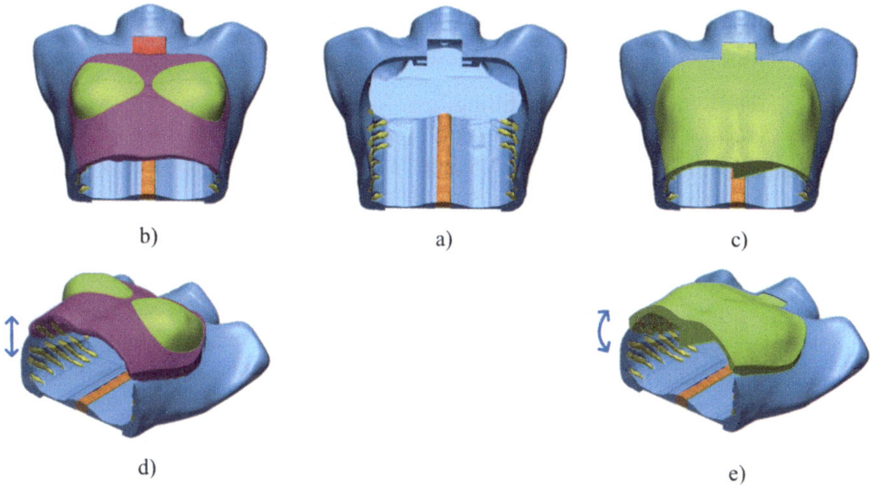

Fig. 3. ADAM ⚥'s and ADAM's different chests.

To make the transition from male to female phantom easier, the cylindrical joint was integrated into the male hatch. The torso was then modified to allow for rotation. Finally, a special plug was created for use with the female hatch, as shown in Fig. 3a. For the movement of the female hatch, the carbon rods (8) shown in Fig. 1 were used as a support on which a holder was modelled to hold the hatch.

2.3 Material Analysis

To produce the different parts of the phantom, a literature search was conducted to understand the current state of the art of 3D printable and non-3D printable materials, with the aim of choosing those with tissue- mimicking properties. The research focused on materials to mimic major anatomical regions not found in ADAM (breast tissue) and on common ones (heart tissue, torso, chest, spine and ribs).

Badiuk et al. [4] tested polyurethane rubber (PMC-780 dry) with borosilicate glass microspheres to replicate soft tissue and showed that the spheres reduced the radiodensity coefficient. Okkalidis et al. [5] evaluated 11 FDM materials to understand how HU varies with part infill and found that 90% infill for PLA produced HU values like adipose tissue. Hatamikia et al. [6] used StoneFil™, a PLA with cement powder, to achieve HU values from −482 to 968 by varying the infill. Talalwa et al. [9] used PORO-LAY elastomeric polymers, which become microporous after immersion in water, showing a correlation between density and HU value.

It should also be considered that breast tissue consists of several parts: skin, fat and glandular tissue. It has been shown [7] that when breast tissue is exposed to energies greater than 50 keV, it can be approximated as a single tissue, namely adipose tissue.

The materials were then tested for their radiological properties, focusing on the HU values obtained from CT scans to ensure that they matched those of real tissue as closely as possible.

As introduced, several materials were tested to faithfully reproduce the HU values of the anatomies involved. The materials tested were:

- PLA samples with 90% gyroidal filling.
- GEL-LAY samples with 15, 50 and 100% gyroidal filling.
- Polyurethane rubber samples with 3.5 and 3% borosilicate nanospheres added.
- StoneFil™ samples with 20, 40, 60, 80 and 100% gyroidal fill.

The 100% filled GEL-LAY sample did not lose its PVA component after 4 days of immersion in water and retained its solid form. The 50% filled sample softened after 3 days but the print layers delaminated. The 15% filled sample lost the PVA and softened without peeling. Only the last sample was subjected to CT. The material analysis showed that the selected materials faithfully reproduced the radiological properties of human tissue. In particular, the Hounsfield Units (HU) measured for the different components of the ADAM ⚥were in line with those of the target tissues, ensuring an accurate simulation of clinical conditions (Table 1).

Table 1. CT Results of Materials.

Material	Infill %	HU
PLA	90	−108
GEL-LAY	15	−665
	50	–
	100	–
Polyurethane rubber	3.5%	−91
	3%	−78
StoneFil™	20	−644
	40	−365
	60	−96
	80	230
	100	692

After obtaining the results of the various materials tested to replicate the tissues under investigation, PLA was chosen as the material for the torso and tissues, while StoneFil™ was chosen for the bone components.

The phantom was printed using the PRUSA MK3S+. The torso was divided into five parts to fit into the print volumes, the female hatch into four parts, as well as the male one. The belly, which can be seen in Fig. 4, having only an aesthetic function, was printed in a four-part division also in PLA.

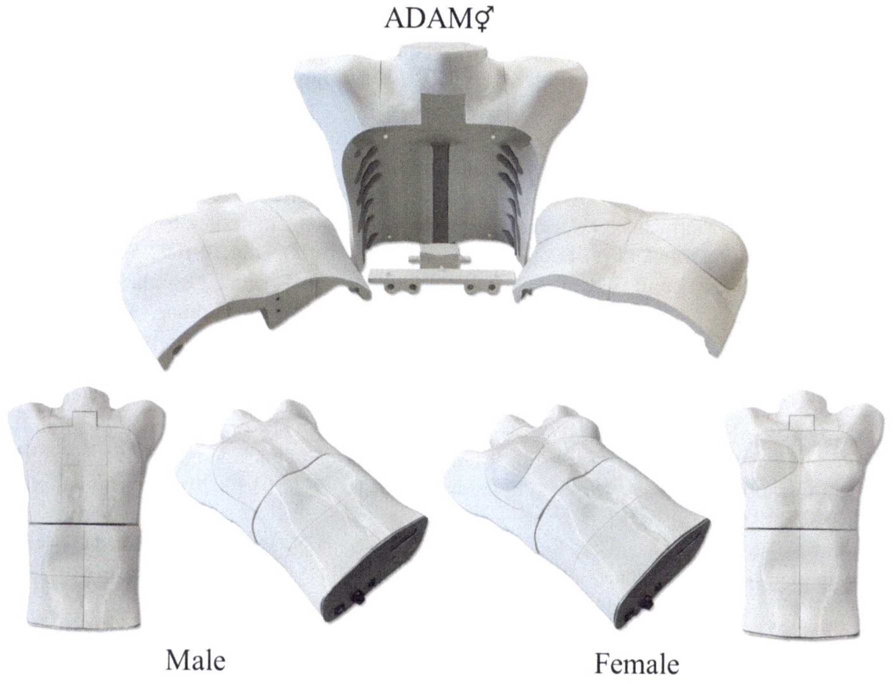

Fig. 4. ADAM ⚥modular Phantom: male version left, female one right.

3 Conclusion

The present study led to the creation of the ADAM ⚥model, a modular phantom based on the previous ADAM, designed to test breast radiotherapy treatments under free breathing and DIBH conditions. The materials chosen, PLA for the torso and StoneFil™ for the bone components, have been shown to accurately replicate the HU values of human tissue, ensuring adequate simulation of real clinical conditions. ADAM ⚥ 's modular design allows a quick transition between male and female configurations, making it suitable for simulating many clinical situations. Furthermore, thanks to the use of the same electromechanical structure, but with different motor applications, it is possible to achieve different movements in a simple way.

These features make ADAM ⚥a more versatile and accurate phantom, ideal for testing and calibrating gating and tracking devices for mobile lesions, improving the accuracy and reliability of radiotherapy treatments.

References

1. Webb, S.: Motion effects in (intensity modulated) radiation therapy: a review. Phys. Med. Biol. **51**, R403–R425 (2006). https://doi.org/10.1088/0031-9155/51/13/R23
2. Pallotta, S., Calusi, S., Foggi, L., Lisci, R., Masi, L., Marrazzo, L., Talamonti, C., Livi, L., Simontacchi, G.: ADAM: a breathing phantom for lung SBRT quality assurance. Phys. Med. **49**, 147–155 (2018). https://doi.org/10.1016/J.EJMP.2017.07.004

3. Pallotta, S., Calusi, S., Masi, L., Talamonti, C., Marrazzo, L., Foggi, L., Casati, M., Livi, L., Simontacchi, G., Desideri, I., Lisci, R.: ADAM phantom to test 4D medical imaging and dose delivery devices. Phys. Med. Biol. **64**, 105002 (2019). https://doi.org/10.1088/1361-6560/AB1A49
4. Badiuk, S.R., Sasaki, D.K., Rickey, D.W.: An anthropomorphic maxillofacial phantom using 3-dimensional printing, polyurethane rubber and epoxy resin for dental imaging and dosimetry. Dentomaxillofac. Radiol. **51** (2022). https://doi.org/10.1259/DMFR.20200323
5. Okkalidis, N., Marinakis, G.: Technical note: accurate replication of soft and bone tissues with 3D printing. Med. Phys. **47**, 2206–2211 (2020). https://doi.org/10.1002/MP.14100
6. Hatamikia, S., Kronreif, G., Unger, A., Oberoi, G., Jaksa, L., Unger, E., Koschitz, S., Gulyas, I., Irnstorfer, N., Buschmann, M., Kettenbach, J., Birkfellner, W., Lorenz, A.: 3D printed patient-specific thorax phantom with realistic heterogenous bone radiopacity using filament printer technology. Z. Med. Phys. **32**, 438–452 (2022). https://doi.org/10.1016/J.ZEMEDI.2022.02.001
7. McGarry, C.K., Grattan, L.J., Ivory, A.M., Leek, F., Liney, G.P., Liu, Y., Miloro, P., Rai, R., Robinson, A.P., Shih, A.J., Zeqiri, B., Clark, C.H.: Tissue mimicking materials for imaging and therapy phantoms: a review. Phys. Med. Biol. **65**, 23TR01 (2020). https://doi.org/10.1088/1361-6560/ABBD17
8. Anwari, V., Lai, A., Ursani, A., Rego, K., Karasfi, B., Sajja, S., Paul, N.: 3D printed CT-based abdominal structure mannequin for enabling research. 3D Printing in Medicine. **6**(1 6), 1–12 (2020). https://doi.org/10.1186/S41205-020-0056-9
9. Talalwa, L., Natour, G., Bauer, A., Drzezga, A., Beer, S.: Radiological characteristics of a new experimental rubber elastomeric polymer used in three-dimensional printing with different infill densities and patterns. J. Phys. Commun. **4**, 125006 (2020). https://doi.org/10.1088/2399-6528/ABD1C3
10. 3D StoneFil - High quality filaments – Formfutura, https://formfutura.com/product/stonefil/

GelMA-Xanthan Gum Hydrogel Bioink Preserving hiPSCs Viability and Proliferation

Virginia Deidda[1], Isabel Ventisentte[1], Marianna Langione[2], José Manuel Pioner[2,3], Caterina Credi[4,5], and Federico Carpi[1](✉)

[1] Department of Industrial Engineering, University of Florence, Florence, Italy
federico.carpi@unifi.it

[2] Department of Experimental and Clinical Medicine, University of Florence, Florence, Italy

[3] Department of Biology, University of Florence, Florence, Italy

[4] European Laboratory for Non-Linear Spectroscopy, University of Florence, Sesto Fiorentino, Italy
caterina.credi@ino.cnr.it

[5] National Institute of Optics – National Research Council, Sesto Fiorentino, Italy

Abstract. Cardiac tissue engineering is currently investigated with a diversity of biofabrication techniques. In particular, 3D bioprinting makes it possible to develop constructs with an organized distribution of cells. Over the years, a wide variety of materials has been used to create environments that emulate the extracellular matrix. Among these, gelatin methacrylate (GelMA)-based bioinks have widely been studied for extrusion-type 3D bioprinting. Several investigations have shown that GelMA at low concentrations ($\leq$5% w/v) can promote cell viability, although limiting 3D printability, owing to poor retention of shape after extrusion. In this work, a new bioink formulation was developed to improve printability, while maintaining a high cell viability. For this purpose, xanthan gum (XG) was added to GelMA. The composite hydrogel had an elastic modulus of ~9 kPa, comparable to typical values of cardiac tissue. 3D printing tests (without cells) showed the printability of the new hydrogel. Both 2D and 3D cellular adhesion and viability tests showed the hydrogel ability to preserve survival and growth of human induced pluripotent stem cells (hiPSCs).

Keywords: Bioink · Stem Cells · Hydrogel · HiPSCs · Cardiac · 3D Printing · Tissue Engineering

1 Introduction

Tissue engineering is continuously targeting the development of new technologies for creating biological tissues [1]. In particular, 3D bioprinting is intensively investigated to fabricate scaffolds capable of supporting cellular functions. However, common strategies typically envisage that cells are seeded onto prefabricated scaffolds; this usually results in a random distribution of cells, which does not reflect the complex hierarchical organization of functional tissues. Therefore, 3D printing of cell-laden bioinks, commonly referred to as bioprinting, is investigated as a viable alternative [2].

© The Author(s), under exclusive license to Springer Nature Switzerland AG 2025
M. Carfagni et al. (Eds.): IDBN 2024, BioSysRob 35, pp. 9–16, 2025.
https://doi.org/10.1007/978-3-031-94626-4_2

A variety of bioinks for bioprinting has been developed, aiming to provide embedded cells with a viscoelastic and hydrated microenvironment that enables nutrients diffusion and extra-cellular matrix remodeling [3]. The choice of the bioink material is crucial, as it regulates the physical and chemical properties of the final printed constructs [4, 5].

Today, extrusion-type bioprinting is broadly implemented with a variety of hydrogels [6–9]. Among these, gelatin functionalized with methacrylate groups (GelMA) has widely been used in extrudable bioink formulations [10]. To ensure the survival and proliferation of cells, an important role is played by the hydrogel stiffness, which should match the targeted tissue's stiffness, as close as possible.

This work focused on cardiac tissue bioprinting, with the goal of developing a novel 3D printable bioink formulation, which showed ability to preserve the viability and proliferation of human induced pluripotent stem cells (hiPSCs).

Studies have found that the stiffness of cardiac muscle tissue is around 8–12 kPa [11] and it is known that GelMA can have elastic moduli of that order at a concentration of ~5% w/v [12–14]. Moreover, it has been shown that, at concentrations up to ~5% w/v, GelMA promotes cell viability [15]. Nevertheless, at such low concentrations, the low viscosity of GelMA results in poor retention of shape after extrusion, i.e. poor printability [13, 16]. Therefore, in order to obtain a printable bioink, it can be useful to combine GelMA with other polymers.

In this work, a new bioink was created by integrating GelMA with xanthan gum, so as to increase the bioink viscosity and ensure retention of shape after extrusion [5, 17]. In particular, we created a composite hydrogel, which combined 5% w/v GelMA with 1.25% w/v xanthan gum. The latter was used at that concentration, according to an empirical optimization process (not described here, for the sake of brevity), to obtain a printable hydrogel that retained its shape after extrusion and an elastic modulus compatible with that of cardiac tissue. Hereinafter, that composition will be identified as GelMA-XG. That bioink was characterized in terms of mechanical properties, 3D printability, and cellular viability and proliferation.

2 Materials and Methods

2.1 Hydrogel Preparation

In order to obtain the GelMA hydrogel, lyophilized GelMA (Sigma Aldrich, USA) was firstly dissolved into either a phosphate buffered saline (PBS) solution or a cell culture medium (depending on the test of interest, as detailed in the following sections), at a concentration of 5% w/v and at a temperature of 40 °C under magnetic stirring, until it was fully dissolved. Then, a photoinitiator (Lithium phenyl-2,4,6-trimethylbenzoylphosphinate – LAP, Sigma Aldrich, USA) was added to the gelatin solution, at a concentration of 0.25% w/v.

The GelMA-XG hydrogel was prepared by adding xanthan gum (XG) (Sigma Aldrich, USA) to the GelMA solution, at a concentration of 1.25% w/v and at a temperature of 40 °C under magnetic stirring. Then, the LAP photoinitiator was added to the GelMA-XG solution, at a concentration of 0.25% w/v.

Finally, the hydrogels were cross-linked at 405 nm, with an exposure time that depended on the sample thickness (as detailed in the following sections).

2.2 Bioink Preparation

The GelMA-XG solution was used to prepare a bioink, by mixing it with variable contents of hiPSCs, depending on the test of interest (as detailed in the following sections) and cross-linking the biomaterial at 405 nm for 15 s.

2.3 Hydrogel Mechanical Characterisation

In order to characterise the hydrogel stiffness, indentation tests were performed as follows. For each type of hydrogel (GelMA and GelMA-XG), 3 mm-thick specimens were obtained by mold casting. The samples were cross-linked for 30 s per side at 405 nm.

In order to quantify the mechanical stiffness of the hydrogel matrix in a fully hydrated state, the samples were first immersed in a phosphate buffered saline (PBS) and brought to swelling equilibrium (i.e., stabilized weight) at room temperature, and then they were subject to indentation tests. The latter were carried out using a bespoke system, consisting of a cylindrical indenter (diameter of 5 mm), attached to a load cell, which was mounted on a micrometric three-axial translation stage.

Indentation steps equal to 5% of the sample thickness were progressively applied. At each step, the force exerted by the specimen in response to the applied indentation was recorded. As a result, the stress-relative indentation curve of the sample was determined.

2.4 Hydrogel Printability Tests

Prior to 3D printing the bioink, the printability of the GelMA-XG hydrogel was evaluated as follows. An extrusion-type bioprinter (BIOX6, Cellink, Sweden) was used to create several horizontal lines, by varying the temperature (20 to 26 °C), speed (5 to 6 mm/s) and pressure (15 to 25 kPa) of extrusion through a 22G conical needle. The samples were cross-linked at 405 nm for 15 s.

2.5 HiPSCs Adhesion and Viability Tests on 2D Cultures on Top of the New Hydrogel

In order to investigate hiPSCs adhesion and viability on 2D cultures, the GelMA-XG and GelMA hydrogels were prepared according to the procedure described above, with the only difference that the dissolution of GelMA was performed within a culture medium (mTeSR, Stem Cell Technologies, USA) instead of PBS.

Hydrogel layers (1–2 mm thick) were created by mold casting and cross-linked at 405 nm for 15 s. The layers' surface was seeded with hiPSCs (approximately ~2 $x10^3$ cell/mm^2), following their chemical dissociation and resuspension in the culture medium (mTeSR, Stem Cell Technologies, USA), which was complemented with a ROCK Inhibitor (RI) (Stem Cell Technologies, USA) at a concentration of 5 μM. Samples were kept in culture at 37 °C and 0.5% CO_2, up to 10 days, and were observed with an optical microscope (EVOS FL auto 2, Thermo Fisher Scientific Inc., USA).

2.6 HiPSCs Survival and Proliferation Tests Within 3D Cultures Inside the New Hydrogel

In order to investigate hiPSCs viability within 3D cultures, the GelMA-XG hydrogel was prepared according to the procedure described in Sect. 3.4. hiPSCs (AICS WTC TNNI1 GFP-labeled control line for ssTnI) were chemically dissociated and resuspended in the mTeSR culture medium, which was complemented with 5 μM RI. The cells were added to the GelMA-XG solution, at a concentration of ~2.3 mln cells/ml, and droplets were created. The samples were cross-linked at 405 nm for 15 s, and were cultured in the mTeSR culture medium with the addition of RI at a concentration of 5 μM at 37 °C and 5% CO_2. Optical microscope images (Olympus CKX53, Life science – Evidence, Japan) were acquired.

3 Results and Discussion

3.1 Hydrogel Stiffness

The results of the indentation tests on the GelMA-XG composite hydrogels are shown in Fig. 1. The elastic modulus of each hydrogel was calculated as the initial slope of a second-order fitting curve for the data. The GelMA hydrogel showed an elastic modulus of ~4 kPa, which was in agreement with the literature [14]. On the other hand, the addition of the xanthan gum led to an increase in stiffness, resulting in an elastic modulus of ~9 kPa, which was comparable with those of cardiac muscle tissue (8–12 kPa) [11].

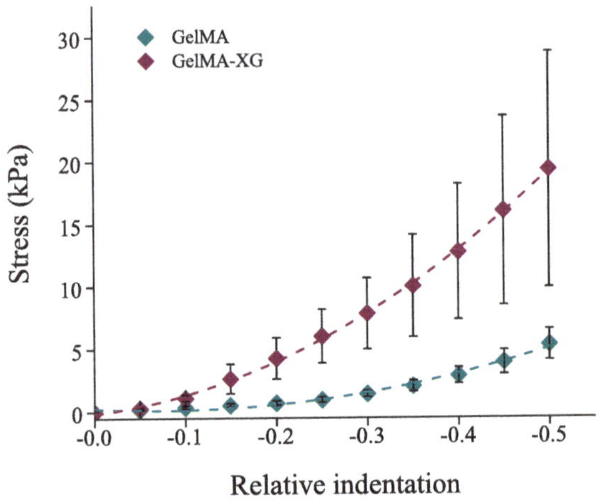

Fig. 1. Stress-relative indentation curve of the GelMA-XG hydrogel, also compared to that of the GelMA formulation. The error bars correspond to the standard deviation among four samples.

3.2 Hydrogel Printability

In order to enable 3D printing, the hydrogel viscosity should be sufficiently low to both ensure a homogeneous distribution of the cells and protect them from excessive shear stresses during extrusion through sub-millimetric needles. On the other hand, the viscosity should also be sufficiently high to ensure that the printed material maintains its shape after extrusion, during the curing process. Therefore, a trade-off between these opposite needs is always necessary.

In this work, a suitable trade-off between ease of extrusion and shape retention was empirically achieved by using the following printing parameters, which were empirically identified with trial-and-errors tests: extrusion temperature: 22.5 °C; extrusion pressure: 17 kPa; extrusion speed: 5 mm/s; extrusion needle size: 22G.

The printability of the GelMa-XG hydrogel using those parameters is shown in Fig. 2, which presents examples of well-defined printed linear structures, having a width of ~700 µm and a thickness of ~400 µm.

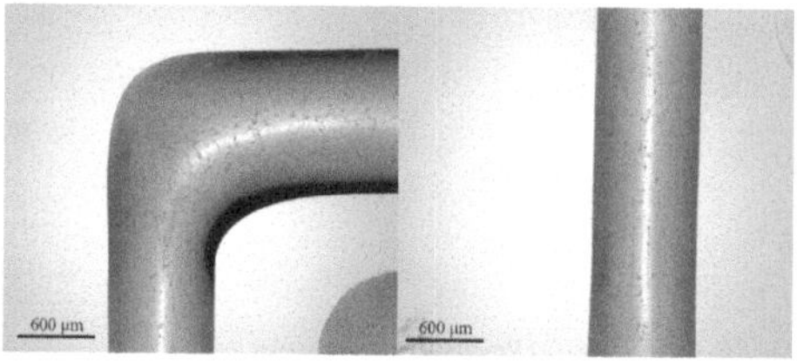

Fig. 2. Optical microscopy images of 3D printed structures made of the GelMA-XG hydrogel. Note that in those preliminary printing tests the hydrogel was also loaded with hiPSCs.

3.3 HiPSCs Adhesion and Viability on 2D Cultures on Top of the New Hydrogel

The result of the adhesion and viability test on 2D cultures up to 9 days after cell seeding on the top of GelMA and GelMA-XG hydrogels are shown in Fig. 3.

From day 4 to 9, cell colonies increased in number and size, proving the ability of both biomaterials to support the proliferation of cells.

3.4 HiPSCs Survival and Proliferation Within 3D Cultures Inside the New Hydrogel

The results of the proliferation test inside the GelMA-XG hydrogel are shown by the images presented in Fig. 4.

They demonstrated the biocompatibility of the hydrogel and its ability to support cell viability. Indeed, the material was able not only to ensure the survival of the cells, but also to support their proliferation.

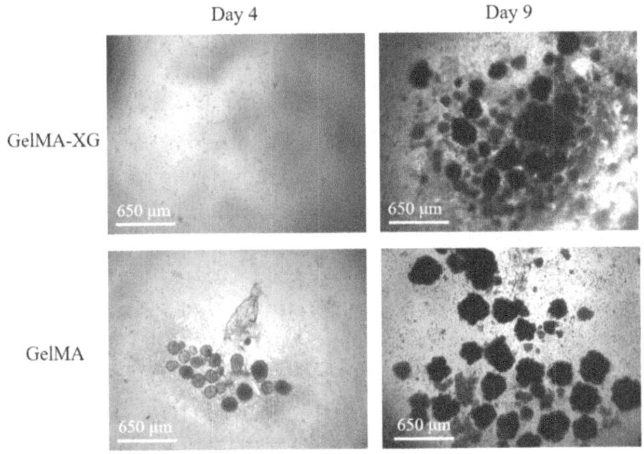

Fig. 3. Light microscopy images of hiPSCs cultured on top of layers of the GelMA-XG and GelMA hydrogels, taken after 4 and 9 days from seeding.

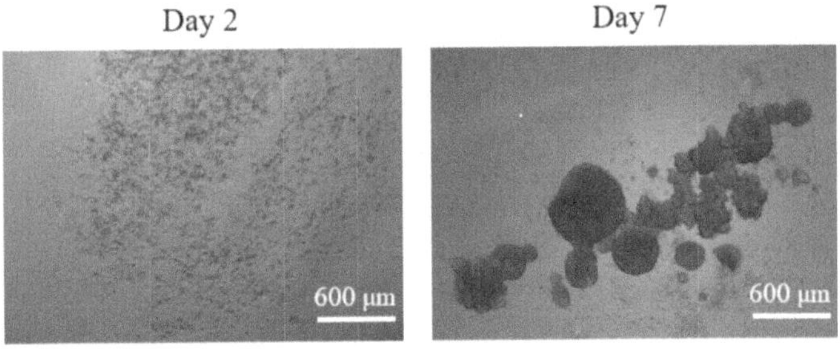

Fig. 4. Optical microscopy images of hiPSCs in culture within the GelMA-XG hydrogel after 2 and 7 days from cross-linking.

4 Conclusions and Future Developments

This work focused on cardiac tissue bioprinting, with the goal of developing a novel bioink formulation allowing for human induced pluripotent stem cells viability and proliferation. Although gelatin methacrylate (GelMA) at low concentrations (<5% w/v) had previously been demonstrated to be usable as a biocompatible material, promoting growth of a diversity of cells, it is also known to limit the printability of structures that retain shape after extrusion, owing to a low viscosity. In this work, to ensure both the printability of the bioink and high cell proliferation, a new GelMA-based bioink was created with the addition of xanthan gum (XG).

The stiffness of the GelMA-XG hydrogel (about 9 kPa) was comparable to that of cardiac tissue. Cellular tests demonstrated the ability of this new bioink to preserve cell viability and proliferation.

In conclusion, the results obtained so far lead us to believe that the developed bioink has potential as a bioprinting material. Future developments of this work will focus on two goals: (i) proving the ability of the bioink to ensure hiPSCs differentiation into cardiomyocytes; (ii) performing systematic bioprinting tests and demonstrating cell viability and proliferation after extrusion.

Acknowledgement. Financial support from the Italian Ministry of University and Research (PNRR project "AGE-IT - Ageing well in an ageing society") is gratefully acknowledged.

References

1. Zorlutuna, P., Annabi, N., Camci-Unal, G., Nikkhah, M., Cha, J.M., Nichol, J.W., et al.: Microfabricated biomaterials for engineering 3D tissues. Adv. Mater. **24**(14), 1782–1804 (2012). https://doi.org/10.1002/adma.201104631
2. Jain, T., Baker, H.B., Gipsov, A., Fisher, J.P., Joy, A., Kaplan, D.S., Isayeva, I.: Impact of cell density on the bioprinting of gelatin methacrylate (GelMA) bioinks. Bioprinting. **22**, e00131 (2021). https://doi.org/10.1016/j.bprint.2021.e00131
3. Kim, J.S., Hong, S., Hwang, C.: Bio-ink materials for 3D bio-printing. J. Int. Soc. Simul. Surg. **3**(2), 49–59 (2016). https://doi.org/10.18204/JISSiS.2016.3.2.049
4. Fan, C., Wang, D.A.: Macroporous hydrogel scaffolds for three-dimensional cell culture and tissue engineering. Tissue Eng. Part B Rev. **23**(5), 451–461 (2017). https://doi.org/10.1089/ten.TEB.2016.0465
5. Iervolino, F., Belgio, B., Bonessa, A., Potere, F., Suriano, R., Boschetti, F., et al.: Versatile and non-cytotoxic GelMA-xanthan gum biomaterial ink for extrusion-based 3D bioprinting. Bioprinting. **31**, e00269 (2023). https://doi.org/10.1016/j.bprint.2023.e00269
6. Liu, W., Heinrich, M.A., Zhou, Y., Akpek, A., Hu, N., Liu, X., et al.: Extrusion bioprinting of shear-thinning gelatin methacryloyl bioinks. Adv. Healthc. Mater. **6**(12), 1601451 (2017). https://doi.org/10.1002/adhm.201601451
7. Skardal, A., Zhang, J., Prestwich, G.D.: Bioprinting vessel-like constructs using hyaluronan hydrogels crosslinked with tetrahedral polyethylene glycol tetracrylates. Biomaterials. **31**(24), 6173–6181 (2010). https://doi.org/10.1016/j.biomaterials.2010.04.045
8. Schöneberg, J., De Lorenzi, F., Theek, B., Blaeser, A., Rommel, D., Kuehne, A.J., et al.: Engineering biofunctional in vitro vessel models using a multilayer bioprinting technique. Sci. Rep. **8**(1), 10430 (2018). https://doi.org/10.1038/s41598-018-28715-0
9. Hong, S., Sycks, D., Chan, H.F., Lin, S., Lopez, G.P., Guilak, F., et al.: 3D printing of highly stretchable and tough hydrogels into complex, cellularized structures. Adv. Mater. (Deerfield Beach, Fla.). **27**(27), 4035 (2015). https://doi.org/10.1002/adma.201501099
10. Tigner, T.J., Rajput, S., Gaharwar, A.K., Alge, D.L.: Comparison of photo cross linkable gelatin derivatives and initiators for three-dimensional extrusion bioprinting. Biomacromolecules. **21**(2), 454–463 (2019). https://doi.org/10.1021/acs.biomac.9b01204
11. Guimarães, C.F., Gasperini, L., Marques, A.P., Reis, R.L.: The stiffness of living tissues and its implications for tissue engineering. Nature Reviews Materials. **5**(5), 351–370 (2020). https://doi.org/10.1038/s41578-019-0169-1
12. Bertassoni, L.E., Cardoso, J.C., Manoharan, V., Cristino, A.L., Bhise, N.S., Araujo, W.A., et al.: Direct-write bioprinting of cell-laden methacrylated gelatin hydrogels. Biofabrication. **6**(2), 024105 (2014). https://doi.org/10.1088/1758-5082/6/2/024105
13. Pepelanova, I., Kruppa, K., Scheper, T., Lavrentieva, A.: Gelatin-methacryloyl (GelMA) hydrogels with defined degree of functionalization as a versatile toolkit for 3D cell culture and extrusion bioprinting. Bioengineering. **5**(3), 55 (2018). https://doi.org/10.3390/bioengineering5030055

14. Zhu, W., Qu, X., Zhu, J., Ma, X., Patel, S., Liu, J., et al.: Direct 3D bioprinting of prevascularized tissue constructs with complex microarchitecture. Biomaterials. **124**, 106–115 (2017). https://doi.org/10.1016/j.biomaterials.2017.01.042
15. Yin, J., Yan, M., Wang, Y., Fu, J., Suo, H.: 3D bioprinting of low-concentration cell-laden gelatin methacrylate (GelMA) bioinks with a two-step cross-linking strategy. ACS Appl. Mater. Interfaces. **10**(8), 6849–6857 (2018). https://doi.org/10.1021/acsami.7b16059
16. Shie, M.Y., Lee, J.J., Ho, C.C., Yen, S.Y., Ng, H.Y., Chen, Y.W.: Effects of gelatin methacrylate bio-ink concentration on mechano-physical properties and human dermal fibroblast behavior. Polymers. **12**(9), 1930 (2020). https://doi.org/10.3390/polym12091930
17. Li, J., Moeinzadeh, S., Kim, C., Pan, C.C., Weale, G., Kim, S., et al.: Development and systematic characterization of GelMA/alginate/PEGDMA/xanthan gum hydrogel bioink system for extrusion bioprinting. Biomaterials. **293**, 121969 (2023). https://doi.org/10.1016/j.biomaterials.2022.121969

Development and Implementation of a Pediatric Simulator for Training in Intraosseous Access Procedure

Antonio Marzola[1], Marta Mencarelli[1](✉), Lorenzo Torzini[1], Marco De Luca[2], Monica Carfagni[1], and Yary Volpe[1]

[1] Department of Industrial Engineering, University of Florence, Florence, Italy
marta.mencarelli@unifi.it

[2] Meyer Children's University Hospital, IRCCS, Florence, Italy

Abstract. Intraosseous access is a critical emergency procedure for introducing fluids and blood components into the systemic circulation, necessitating thorough training for healthcare personnel. In collaboration with Meyer Children's University Hospital IRCCS, the Department of Industrial Engineering at the University of Florence has developed a high-fidelity, low-cost pediatric simulator specifically for intraosseous access in the proximal tibia. Using Materialise Mimics 26® software, anonymized tomographic images of a 5-year-old patient were processed to obtain the three-dimensional geometry of relevant anatomical structures: tibia, fibula, distal femur, muscles, and skin. These models were further refined with Geomagic DesignX® for the final simulator design.

The simulator features silicone replicas for soft tissues, with rigid structures and silicone casting molds produced through Fused Deposition Modeling. Preliminary tests were conducted to select suitable materials, which were then evaluated by specialized medical personnel using the EZ-IO® device to ensure accurate haptic feedback. The resulting simulator provides a realistic platform for intraosseous access training, enabling healthcare practitioners to enhance their procedural skills and ensure patient safety.

Keywords: additive manufacturing · pediatric mannequin · intraosseous access

1 Introduction

Intraosseous (IO) access is a vital procedure in emergency situations, enabling the rapid infusion of fluids and blood components directly into the systemic circulation when traditional venous routes are not easily accessible [1]. This technique is especially crucial in pediatric patients, where immediate response is essential and the conditions are often unstable [2, 3]. IO access is performed using specialized devices that typically feature powered or manual insertion mechanisms and specialized needles designed to penetrate the bone and deliver fluids and medications directly into the bone marrow. Performing this procedure requires high competence from healthcare personnel [3, 4]. The difficulty of this procedure is amplified in pediatric patients due to their smaller size and unique anatomical characteristics.

© The Author(s), under exclusive license to Springer Nature Switzerland AG 2025
M. Carfagni et al. (Eds.): IDBN 2024, BioSysRob 35, pp. 17–24, 2025.
https://doi.org/10.1007/978-3-031-94626-4_3

Currently, several simulators on the market are specifically designed to train the IO access skill. However, practitioners often complain about these simulators' lack of anatomical realism. Additionally, the high initial cost of some models (€500–€1500) can be a significant obstacle for many healthcare institutions. Over the past few years, thorough research has been carried out by a research group aiming to develop new simulators to improve upon commercially available devices by incorporating expert user feedback. Between 2020 and 2022, the research team developed a series of innovative simulators for tibial intraosseous access to enhance medical training in this area for adult and pediatric patients [5–8]. The materials used included PLA and TPE for the 3D-printed parts, and various types of silicone to simulate soft tissue. Production costs ranged from €37 to €52 per unit, making them relatively affordable for training purposes. Each simulator proposed by the authors was rigorously tested by healthcare providers, providing effective results.

However, an examination of the collected data revealed several limitations:

- low anatomical accuracy, posing challenges for operators in identifying the access area.
- limited effectiveness of the pediatric model, since it was obtained by scaling down from an adult model by 60%.
- movement between the skin and subcutaneous layers feels unnatural.

Both the state-of-the-art review and market analysis indicate a deficiency in effective pediatric intraosseous access simulators tailored for patients under 5 years of age. To address this challenge and enhance the training of medical personnel effectively, the Department of Industrial Engineering at the University of Florence and Meyer University Hospital IRCCS collaborated within the T3Ddy laboratory to develop a high-fidelity, low-cost pediatric simulator for intraosseous access in the proximal tibia. The specific access point of the proximal tibia was chosen because it is the most common insertion point for patients aged below 5 years, since it is easier to locate and penetrate compared to other options. Drawing insights from both a comprehensive state-of-the-art review and feedback from practitioners training on existing commercial simulators, this research initiative aims to deliver an innovative and reliable tool to train the procedural skills of healthcare professionals for the IO access procedure.

2 Materials and Methods

The development process of the pediatric simulator for IO access was guided by a rigorous methodology that involved several phases, from the collection of anatomical data to the selection of materials and the creation of the first prototype.

2.1 Anatomical Data Collection

The anatomical features of the simulator were based on anonymized tomographic images of the left leg of a 5-year-old patient. These images were processed using Materialise Mimics 26® software to extract the three-dimensional geometry of the anatomical structures relevant to the simulation. The structures of interests include:

- **Bones**: Tibia, fibula, distal femur, patella.
- **Internal soft tissues**: All structures except bones and skin.
- **Skin**.

Since the computed tomography (CT) scan did not include the ankle and foot, the skin and internal soft tissues were derived from a digital representation of a 5-year-old model using the MakeHuman™ software. This software provided a 3D model of the external surface of a human body in STL format. The left leg was isolated from the complete model, properly scaled, and aligned to match the skin segmented from the CT scan. The skin was defined by thickening inward from the external surface of the synthetic model. The thickness of the skin was set to 2.5 mm to balance realism and technological feasibility, considering that the skin would be produced through silicone casting and a thinner layer could pose challenges during the casting process. The internal soft tissues were obtained by subtracting the skin and bones from the leg model. Treating all soft tissues except the skin as a single volume simplifies the model without sacrificing functionality. The entire model, including the bones, skin, and internal soft tissue, was delimited proximally by a plane defined perpendicular to the femur axis and positioned 50 mm above the distal femoral physis, ensuring the knee joint was included in the simulator. As a result, the entire leg was approximately 270 mm long.

2.2 Layout Definition and Design

To achieve the desired specifications, the simulator was divided into four macro-regions, each with distinct functions influencing the design of the simulator's components:

- **Simulation Region**: This section is critical for ensuring correct feedback to the IO access device. It includes the anatomy near the insertion point, focusing on the proximal tibial access in this application. The accuracy of both device feedback and anatomical features is paramount in this region.
- **Reference Anatomy Region**: This group includes components that are not directly involved in the procedure but serve as anatomical references. While they do not need to provide correct feedback for the device, they must offer appropriate anatomical accuracy.
- **Realism Enhancement Region**: This section provides additional anatomical references to enhance the simulation's realism. Although the anatomical accuracy and functionality of these regions can be approximated, they effectively contribute to the overall realism.
- **Ancillary Support Region**: These parts support the simulator's functionality, including the support system.

The functions of each macro-region influenced the design choices for each component of the simulator. The components of the simulator are as follows:

Cartridge: This is a critical component, the only one included in the Simulation Region, since it includes the actual working area. Given the critical importance of both device feedback and anatomical accuracy in this region, the shape of the hard tissues closely replicates the bone models segmented from CT scans. According to [9], proximal tibial insertion is typically performed at the center of the anteromedial surface of the tibia, approximately 1–2 cm below the inferior pole of the tibial tuberosity. Based on this, the Cartridge includes the tibial bone and internal soft tissue ranging between 10 mm and 40 mm below the tibial plateau. The other dimensions were determined by maximizing the working area while ensuring effective housing and mechanical resistance to the surrounding bone. The back surface of the cartridge (the interface between the cartridge and the remaining tibia) was shaped to ensure a minimum thickness of 5 mm in the central sagittal plane of the tibia. This surface, along with two planes perpendicular to the tibial axis located at the upper and lower limits of the considered volume, was used to delineate the Cartridge from the remaining part of the tibia. Two profiles were added to the upper and lower parts of the cartridge (Fig. 1) to serve as a guide for the insertion of the component and to hold it in place, avoiding rotation or translation during simulation activities. Since the final model of the Cartridge has to include also the internal soft tissues, the same features used to delineate the bone boundaries were applied to delineate the regions of the soft tissues (Fig. 1c). To optimize durability and minimize costs, the Cartridge was designed as a unique replaceable component. This ensures the simulator's long-term effectiveness by allowing worn cartridges to be easily replaced with new ones.

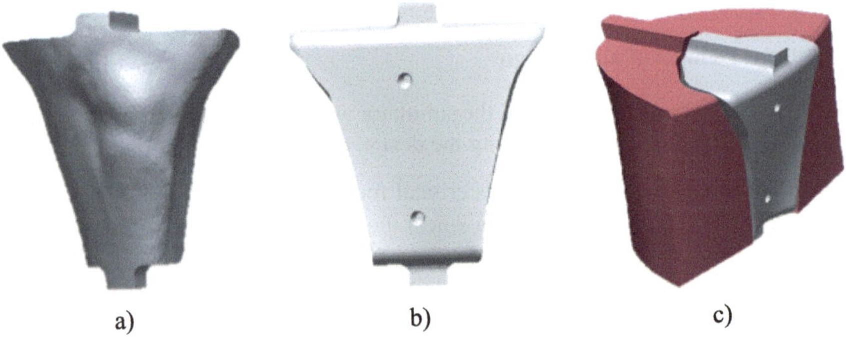

Fig. 1. a) frontal and b) back views of the bone component of the Cartridge, and c) the complete Cartridge with the internal soft tissue included

Skin: The entire skin model was included in the Realism Enhancement Region, as it provides the external interface of the simulator and is crucial for enhancing realism, while its contribution to the IO device feedback proved to be negligible. The skin was designed as a single component to be fitted like a sock over the internal components. This design increases overall realism and ensures that the insertion zone is not directly identifiable, forcing the user to identify the correct working point using anatomical references. However, this approach requires the skin to be removed and replaced along with the exhausted cartridge.

Core: The Core includes all three regions outside the Simulation Region (Fig. 2). The Reference Anatomy Region, closer to the insertion point, includes the knee joint, head of the fibula, and upper proximal third of the tibia. These features provide anatomical references to assist trainees in identifying the correct insertion point. Therefore, it is crucial that all anatomical features within this group ensure an accurate representation, even if they do not need to provide specific device feedback. The Reference Anatomy Region of the Core also include the knee joint, a fundamental feature to locate the correct insertion point. However, precise replication of the knee's full kinematics is unnecessary. The knee joint was then simplified as a simple hinge, by considering the femoral condyles and tibial plateau as cylindrical sliding surfaces; the hinge mechanism was designed to be coaxial with the axis of this kinematic pair. This hinge design does not affect the anatomical reference, as the frontal part of the hinge is completely covered by the patella. The lower limit of the Reference Anatomy Region of the Core was defined 25 mm below the lower limit of the Cartridge. Beyond this point, robust anatomical references are unnecessary, and the Core's Realism Enhancement region starts. In this region, the internal soft tissues and bones were designed as a single volume to maximize practicality and cost-effectiveness without compromising realism. The upper part of the Core is, on the contrary, an Ancillary Support Region, since it integrates a fixing interface; this interface is meant to connect with another component, called the Joint. The Core features also the housing for the Cartridge. The housing was enlarged by 0.2 mm to ensure the convenient insertion of the Cartridge.

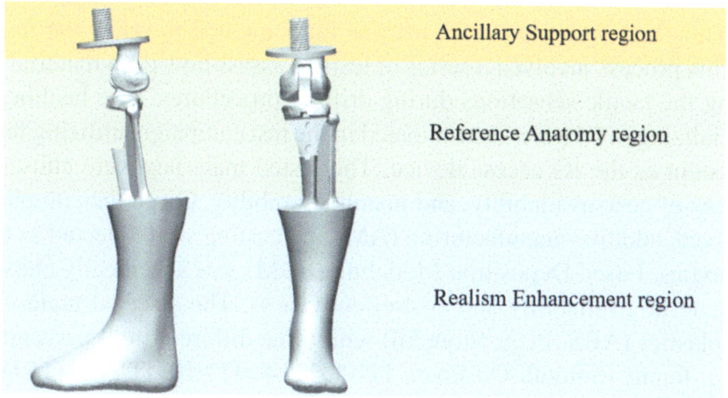

Fig. 2. The Core including all three regions outside the Simulation Region.

Internal soft tissues: Since the distal part of the internal soft tissues was included in the Core, only the proximal part needed to be designed as an independent component. The plane used to divide the Realism Enhancement region from the Reference Anatomy region in the Core was used here to isolate the proximal part of the internal soft tissues. As with the Core, the cartridge volume was cut out from the soft tissues, and the resulting housing was enlarged by 0.2 mm to ensure the proper insertion of the cartridge.

Joint: The Joint is a component included in the Ancillary Support region. It features a hinge at the farthest part from the leg to simulate the hip joint, with an interface on the opposite end to connect the rest of the simulator. A section of this connection is provided in Fig. 3. Along with the Joint, the Ancillary Support Region includes a base to support the device. The Joint is connected to the base by two bolts. This design ensures the simulator is stable during use.

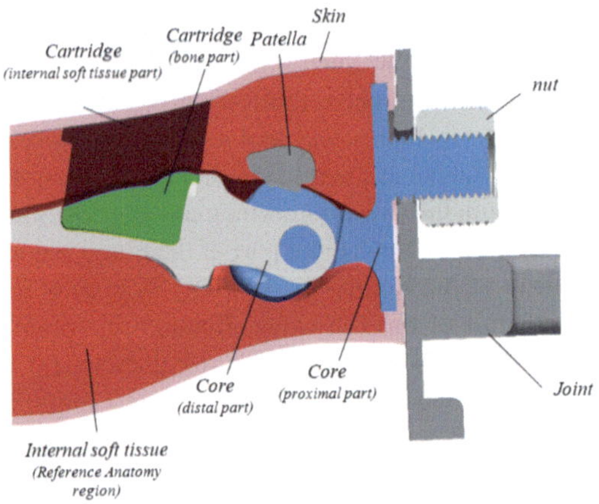

Fig. 3. A section of the proximal region.

2.3 Materials Selection

To ensure realism, a rigorous procedure was implemented to select the most suitable materials. This process involved a series of tests to assess how each material performed in replicating the tactile sensations during drilling procedures. Five healthcare professionals specializing in IO access participated in the test campaign, utilizing the ARROW EZ-IO® System as the IO access device. The tested materials were chosen based on considerations of cost, availability, and manufacturability. Given the complexity of the shapes involved, additive manufacturing (AM) and casting were selected as the fabrication technologies. Fused Deposition Modeling (FDM) was specifically chosen for AM due to its material availability and cost-effectiveness. The selected material included 3D printed plastics (ABS, PLA, Stonefill) with three different infill percentages (40%, 70%, 100%), foams (Smooth-On Foam IT!-5, IT!-8, IT!-10, IT!-15, IT!-IV), a fiber-filled resin (Smooth-On Plasti-Paste II), an ultra-low viscosity casting resin (Smooth-On SmoothCast), and a urethane elastomer (Smooth-On Simpact 85A). Consequently, the test campaign encompassed seventeen specimens, each designed in a cylindrical shape with a radius of 50 mm. PLA and Stonefill with an infill of 40% were evaluated as the best choices, with PLA chosen for its cost-effectiveness and suitable performance. Differently from the hard tissues, soft material selection drew on prior successful experiences [10]; this selection has been then confirmed by the tests. In particular, EcoFlex 00-10 silicone was selected for internal soft tissues, and DragonSkin 30 silicone for the Skin.

2.4 Fabrication

As mentioned in the previous section, the simulator was manufactured using two distinct technologies: FDM for the rigid components and silicone casting for the skin and internal soft tissues. For the non-interchangeable parts (Joint and Core), which do not need to provide adequate device feedback, ABS was chosen due to its superior mechanical strength, despite higher costs, and printed using the Stratasys F370 printer. The interchangeable Cartridge was printed in PLA using the Original Prusa i3 M3KS+ printer. FDM technology was also employed to fabricate the casting molds, utilizing PLA due to its cost-effectiveness.

In terms of costs, the total price is estimated at approximately €128 excluding molds. This includes approximately €2 for the Cartridge, €9 for the skin, and €115 for the remaining components that do not require replacement. The cost of the molds totals around €18.

Figure 4 shows the assembly of the simulator.

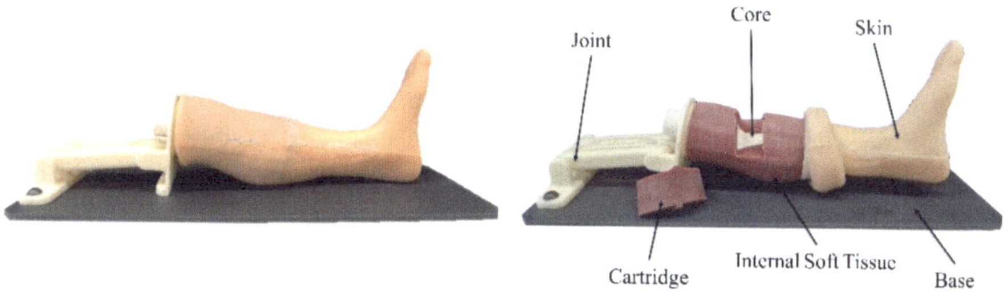

Fig. 4. The fully-assembled simulator (left), and with the uncovered proximal region (right).

2.5 Validation

A series of tests were conducted on the produced simulator to assess its efficiency and realism. Specialized staff at Meyer Children's Hospital used the EZ-IO® device to perform IO accesses on the simulator, evaluating tactile feedback, overall realism, and usability. The results indicated a generally positive user experience, suggesting that the core concept and certain features hold significant promise.

However, the evaluation also identified areas for improvement, primarily:

- Unnatural movements of the skin, which tended to twist with the needle upon full insertion during drilling.
- Difficulty in identifying anatomical references through the internal soft tissue and overlying skin.

Addressing this feedback is crucial for enhancing the simulator to better align with the needs of medical personnel.

3 Discussion and Conclusion

The results of this study demonstrate the effectiveness of the proposed pediatric simulator for intraosseous access. The selection of appropriate materials and careful design has allowed the creation of a high-fidelity simulator capable of faithfully reproducing anatomical characteristics and tactile sensations associated with the real procedure.

These findings confirm the effectiveness and utility of the developed pediatric simulator in enhancing the procedural skills of healthcare personnel and promoting the safety and care of pediatric patients during medical emergencies. They also serve as a valuable starting point for refining the final version of the simulator based on feedback received from physicians at Meyer's Children's Hospital.

Despite the promising development of the pediatric intraosseous access simulator, several limitations need to be addressed to enhance its realism and usability, as indicated by the test results. Firstly, the skin tends to detach from the cartridge, highlighting the need for a redesign that incorporates the skin layer within the cartridge. Although this solution may compromise realism by making cartridge boundaries visible, it addresses a primary issue noted in related literature [5–8].

Secondly, to improve anatomical referencing, the bones need to be scaled to enhance their dimensions while maintaining the volume of the entire leg.

In conclusion, while the developed pediatric simulator offers valuable training opportunities, addressing these limitations is crucial for its advancement. Future development will primarily focus on improving the cartridge to prevent skin detachment and twisting, enhancing the effectiveness of the Reference Anatomy region by refining bone dimensions. These enhancements will not only improve the simulator's realism and functionality but also facilitate its broader adoption in medical training programs.

References

1. Luck, R.P., Haines, C., Mull, C.C.: Intraosseous access. J. Emerg. Med. **39**(4), 468–475 (2010)
2. Sá, R.A., Melo, C.L., Dantas, R.B., Delfim, L.V.: Vascular access through the intraosseous route in pediatric emergencies. Rev. Bras. Ter. Intensiva. **24**(4), 407–414 (2012)
3. Joerck, C., Wilkinson, R., Angiti, R.R., Lutz, T., et al.: Use of intraosseous access in neonatal and pediatric retrieval-neonatal and Pediatric Emergency Transfer Service, New South Wales. Pediatr. Emerg. Care. **39**(11), 853–857 (2023)
4. Engle, W.A.: Intraosseous access for administration of medications in neonates. Clin. Perinatol. **33**, 161–168 (2006)
5. Engelbrecht, R., Patey, C., Dubrowski, A., et al.: Development and evaluation of a 3D-printed adult proximal tibia model for simulation training in intraosseous access. Cureus. **12**(12) (2020)
6. Wade, R.E., McCullum, B., Patey, C., et al.: Development and evaluation of a three-dimensional-printed pediatric intraosseous infusion simulator to enhance medical training. Cureus. **14**(1) (2022)
7. Sivanathan, M., Yanguez, F.L., Joshi, S., et al.: Development of simple and advanced adult proximal tibia simulators for a decentralized simulation-based education model to teach paramedics-in-training the intraosseous infusion procedure. Cureus. **14**(10) (2022)
8. Sivanathan, M., Micallef, J., Clarke, K.M., et al.: The development and initial end-point user feedback of a 3D-printed adult proximal tibia IO simulator. Cureus. **14**(5) (2022)
9. Boussicault, G., Ollivier, M., Allard, L., Riguier, B.: Accessi venosi nel neonato, nel lattante e nel bambino. EMC - Anestesia-Rianimazione. **23**(1), 1–24 (2018)
10. Santarelli, C., Puggelli, L., Carfagni, M., Governi, L.: Preliminary study of a high-fidelity simulator for the management of paediatric tracheal pathologies. In: Rizzi, C., Campana, F., Bici, M., Gherardini, F., Ingrassia, T., Cicconi, P. (eds.) Design Tools and Methods in Industrial Engineering II. ADM 2021 Lecture Notes in Mechanical Engineering. Springer, Cham (2022)

Prototyping and Testing of a Novel Simulator of Congenital Heart Diseases for Hands-on Surgical Training

Michele Bertolini[1(✉)], Alessandro Maria Marchini[1], Neel Murthy[2], Francesco De Gaetano[1], Luke Smith[3], Mirjana Cvetkovic[3], Giorgio Colombo[1], and Claudio Capelli[2]

[1] Politecnico di Milano, Milan, Italy
michele.bertolini@polimi.it
[2] University College London, London, UK
[3] Great Ormond Street Hospital, London, UK

Abstract. Congenital heart diseases (CHD) are structural birth abnormalities of the heart, requiring complex surgical treatment due to anatomical variability and the small size of the affected structures. Hands-on surgical training is essential, and recent advancements in 3D printing have enabled the creation of detailed heart models. However, 3D printed models often lack the mechanical properties of real human tissue. Silicone offers a promising alternative, due to its varied hardness levels and cost-effectiveness. This study aims to propose and test a novel simulator for CHD repair procedures, with a silicone heart model. A pediatric heart with a ventricular septal defect (VSD) was considered and a simulation environment was designed based on the dimensions and proportions of an infant's chest, incorporating a modular platform with a rib cage. A silicone heart was successfully manufactured via a single-piece injection molding procedure. The process used a soluble internal material and a multi-part external shell. The full simulator setup was evaluated by 6 surgeons to address anatomical accuracy and suitability in surgical training of CHD. Despite some limitations and drawbacks were highlighted, the presented solution was found to be an overall useful tool for enhancing surgical training in CHD repair.

Keywords: Congenital Heart Diseases (CHD) · Ventricular Septal Defect (VSD) · Simulation-based training · 3D printing · Silicone molding

1 Introduction

Congenital heart diseases (CHD) represent a group of structural birth abnormalities in the heart. They result from errors in the development of the heart during the first 6 weeks of pregnancy. Most CHD require surgical treatment in the first years of life, and this type of surgery is extremely complex because of the pathological anatomy and the small size of the cardiovascular structures. For these reasons, specific training like hands-on surgical training is crucial to develop surgical skills [1, 2].

© The Author(s), under exclusive license to Springer Nature Switzerland AG 2025
M. Carfagni et al. (Eds.): IDBN 2024, BioSysRob 35, pp. 25–32, 2025.
https://doi.org/10.1007/978-3-031-94626-4_4

The diffusion of 3D printing technology has allowed to directly produce highly detailed heart models, typically obtained from segmentation of CT or MRI scans [3, 4]. The main limitation of this technology when used for complex surgical training comes with the material properties. Indeed, even if it is possible to print soft materials, also with the possibility of creating blends with different shore hardness, their mechanical response is still far from the real behavior of human tissue. One of the main limitations of 3D printed models is the lack of tactile similarity to live tissue [5], together with a tear resistance experienced as insufficient.

An alternative solution relies on the use of silicone. Silicone has the great advantage of being cheaper than commercial resins and being available in a wide range of hardness levels, which guarantees a wide variety of mechanical properties. However, nowadays silicones are generally not directly 3D printable, and the most widespread manufacturing technology is injection molding. Peel et al. [6] showed that the silicone molded heart models provide excellent physical properties for surgical simulation and are extremely resistant to traction forces, whereas the 3D printed heart models can be easily torn during steps like suturing. As a minus, molding is more labor-intensive, requiring the production of multiple molds and silicone pouring. Two main approaches have been found in the literature to manufacture molded silicone heart models. The first solution was proposed by [6] and relies on a "parting and assembly" procedure. Different portions of the heart are produced with different molding processes, and they are then glued together. Another possibility is to print the external shells and the internal core with two different materials, as reported by Thielen et al. [7] and Mattus et al. [8]. In both these cases, the internal part was printed in a soluble material, eliminating the extraction problems.

Creating a significant simulation experience depends not only on the realism of the structure and material of the heart, but also on the entire design of the simulation platform, which should also provide good physical stability to the heart model during the training. Moreover, a good simulator should be modular, allowing different types of heart models to be mounted on it, making it suitable for different types of CHD procedures. An interesting design was proposed by Peel et al. [9]. Their simulator comprehends the operating table simulator, the suture retraction disk (for fixing the thread during the simulation), and the pediatric chest wall cavity.

The aim of this work is to develop and test a new proof of concept of simulator suitable for specialized CHD training of pediatric surgeons, investigating a ventricular septal defect (VSD) case. Starting from the rationale adopted by Peel et al., we aimed to propose a new training solution, integrating ribs and a different fixation system for the heart, together with a redesigned manufacturing strategy for silicone. In Sect. 2, the design and manufacturing procedure for both the silicone heart model and the simulation environment will be presented, together with a description of the methodology used to collect clinical feedback. In Sect. 3, manufacturing results will be shown, as well as aggregated results coming from questionnaires filled by surgeons who tested the simulator. Eventually, Section 4 will analyze the pros and cons of our design and manufacturing approach, and discuss limitations and possible future developments.

2 Materials and Methods

A 3D heart model with a muscular VSD located in the distal part of the ventricles, far from the valves, was selected, freely downloaded from *Sketchfab* online platform [10] and adapted. This choice is related to the relative easiness of manufacturing and operating on this kind of defect. In particular, the original 3D model was edited in 3-Matic environment (Leuven, Belgium), to extend descending aorta and pulmonary arteries, while structures of the heart not necessary for the procedure and complex to be manufactured, like valves, were removed. The external surface was smoothed, and the mesh was refined and fixed to have a watertight printable geometry (Fig. 1).

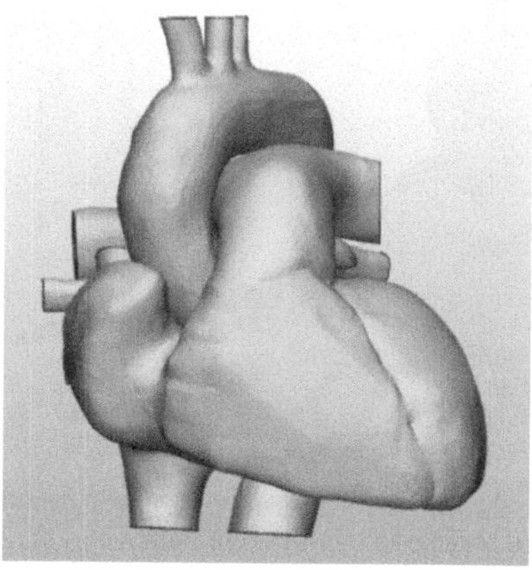

Fig. 1. 3D model of the heart after post-processing in 3-Matic.

Starting from this anatomy, a proper mold was designed, by means of 3-Matic software. The idea was to obtain the heart in a single casting process. To do this, an external shell and an internal core were used. From the 3D model of the heart, two separate surfaces were derived. The internal surface is the one of the blood pool, which is needed to obtain the heart mold cavity. The external surface of the heart is required to create the shell of the mold, by imposing a constant thickness to it. Then, cut planes to partition the external shell and the blood pool in more parts were identified. The external shell was split into 4 parts and the blood pool into 3 parts. In the end, a pin-and-hole system was designed to guarantee the correct alignment and assembly of parts, while air vents and a nozzle for the injection of silicone were added to the model.

Mold manufacturing was performed by means of 3D printing process. ABS filament was used for the blood pool, with an Ultimaker S3 printer (Ultimaker, Utrecht, NL). This choice was related to the good solubility of ABS in acetone, which should allow us to remove the blood pool from the mold. A standard White Resin by Formlabs (Somerville, USA) was used for the external shell, to have a good external finish.

The mold was then assembled, to allow the silicone to be poured in. Dragon Skin™ silicone Shore A 20 (Smooth-On, Macungie, USA) was used. This choice resulted from a previous comparison, in which 2 silicones (shore A 20 and 33, respectively) and a 3D printed resin (shore A 40) were qualitatively evaluated by 8 surgeons, regarding aspects such as palpation, extensibility, cutting ability and tearing. The silicone Shore A 20 proved to be the best option.

Selected silicone was injected through a dispensing gun for two-component silicone into the sealed shell to solidify. Once the silicone set, after 4 h, the shell was removed and the rest was placed in acetone for a couple of days to dissolve the ABS blood pool (Fig. 2).

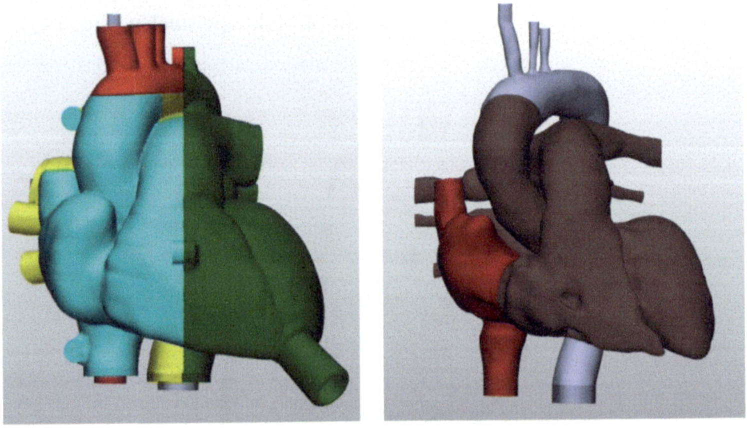

Fig. 2. Heart model for silicone injection molding: on the left, the external shell, on the right, the internal core. Different colors correspond to different parts.

To replicate a realistic simulation environment, a new platform was designed. In 3-Matic software, a rib cage model was imported from *Sketchfab* [11], scaled, and divided into 2 portions, creating a sternum aperture to mimic the surgical chest cut done by surgeons in the classical open-chest procedure. A geometrical matching of each rib head with the platform was designed for assembly. Finally, connections between the platform and the heart model were designed (Fig. 3, colored in blue), exploiting vessels geometry through shape fits, in particular pulmonary arteries, the inferior and the superior vena cava.

ABS was used as platform and rib cage material, printed with an Ultimaker S3, while Formlabs White Resin was used for connectors, for which dimensional accuracy is relevant.

The novel simulator for hands-on surgical training was evaluated by 6 cardio-thoracic pediatric surgeons. A survey was designed to assess the accuracy of the morphology of the heart and the realism and usability of the simulator setup for surgical training. Participants assessed the proposed solution, according to a 5-point Likert scale with binary yes/no questions, grouped into four sections.

- Section 1 compared the digital model on-screen to the manufactured heart in terms of external surface detail, spatial relations of the great vessels, size of the aorta, pulmonary vessels and heart chambers.

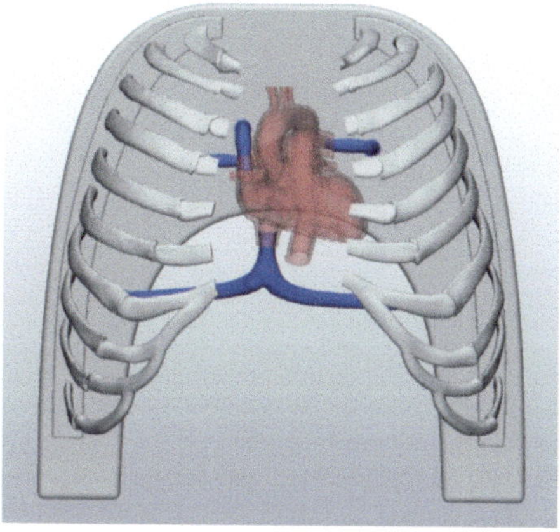

Fig. 3. 3D virtual model of the full simulation platform.

- Section 2 covered the thicknesses of each vessel, atrial and ventricular free walls, directly comparing them to infant's hearts as seen on the operating table. Participants were tasked with incising either the right atrium or ventricle to evaluate the suitability of patching the apical muscular VSD.
- Section 3 evaluated the realism of the material.
- Section 4 was to determine if the setup could be used for teaching students or trainees in basic CHD surgical interventions, rating the 3D-printed ribcage and the setup for overall suitability for training.

3 Results

The whole manufacturing process took our team 5 days to produce a single infant-sized heart ready for the study, with most of the time taken to print blood pools and shells and wait for the silicone to set. After the curing time, the external shell was removed with pliers and the silicone heart, with the ABS core still inside, was put in an acetone bath. After a couple of days, ABS degraded and it was possible to completely remove it. The resulting silicone model is shown in Fig. 4.

The manufacturing of the simulation environment was straightforward, simply relying on 3D printing. After prints postprocessing, it has been proceeded to the assembly phase, to get the result reported in Fig. 5. The assembly phase turned out to be easy, with a stable fixation for the heart.

Coming to the testing phase aggregated feedback, even if preliminary, it is interesting to notice that 5 out of 6 participants determined the spatial relationships between the aorta, pulmonary vessels and their respective chambers were correct. Questionnaires revealed that the right atrium and left ventricle were the most accurate in terms of size, and the left pulmonary artery furthest from accurate. The thickness of the right pulmonary artery and the right atrial free walls were the most accurately represented, closely resembling that of an infant. In contrast, the thickness of the left atrial free wall was the least accurate. Rating the material, extensibility was the most similar aspect to

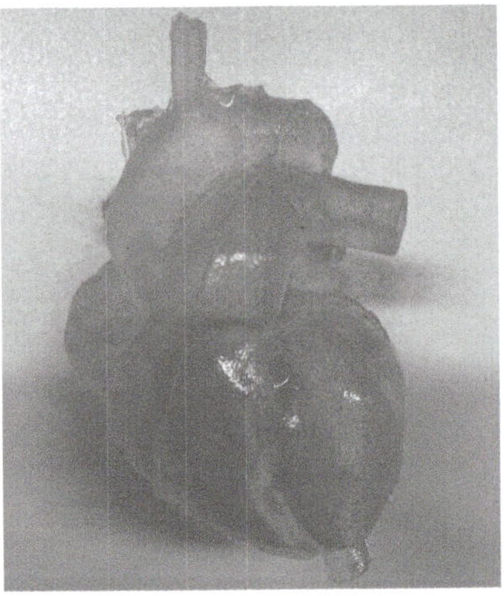

Fig. 4. Silicone heart model obtained through the injection molding process.

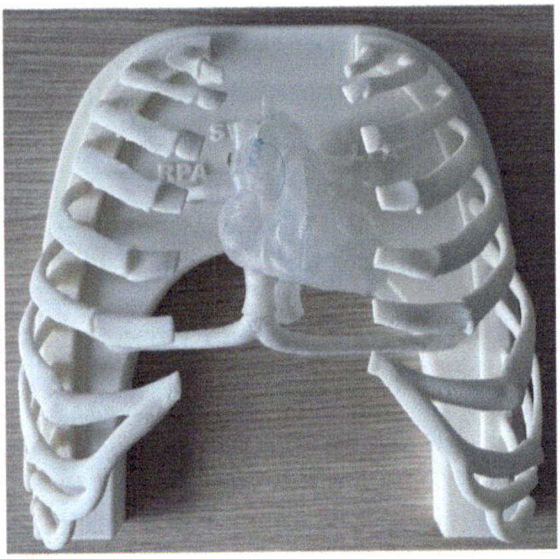

Fig. 5. The assembled physical simulator.

human tissue, whilst tearing along the suture was the least similar, because too low with respect to the real case. The ribcage in the simulator setup was rated as too brittle by all 6 participants, with none of them scoring it as similar to human ribs. 2 participants thought access to the heart with the ribs presented close to a real procedure, but 2 others also found it more difficult.

4 Conclusion and Discussion

In this work, an innovative approach to producing silicone hearts using a single molding technique was successfully implemented, achieving satisfying results in anatomical accuracy. A proper simulation environment was also designed, adding modular features to guarantee the exchangeability of components. The full simulator setup was then evaluated for its anatomical accuracy and suitability in surgical training of a CHD repair procedure.

A notable achievement of this work is the development of a single-casting process for a silicone heart model suitable for surgical simulation. This innovation surpasses previous methods documented in the literature, which involved printing and assembling different parts of the heart. Additionally, a new design of the simulation platform has been conceived, with a modular rib cage. The customized connection between the platform and the specific heart model ensures a stable and realistic setup, improving upon previous methods that used a simple pillar for attachment.

A relevant limitation of this preliminary work is related to the high effort, together with the low repeatability, in producing a heart model with this process. It can be considered a handcrafted procedure, requiring several iterations to get a satisfactory result. For example, scotch tape was used to prevent liquid silicone from leaking out of the mold. Moreover, the external shell, which ideally should have been reusable, had to be broken away with pliers, because it could not be effectively detached. Another challenge encountered was the cleaning phase, with the internal core that did not fully dissolve, due to the complexity of the geometry. Manual removal of undissolved parts was necessary. Despite this, the manufacturing phase, though complex and requiring several attempts, ultimately succeeded in producing a quite accurate silicone heart model.

Future developments could include incorporating additional organs relevant to surgical procedures, together with the creation of a library of segmented CHD models, like for TGA or DORV, to be integrated into the simulation platform. Indeed, even if this study investigated a VSD, the modular concept of the simulator allows the trainee to easily replace the pediatric heart model and so the type of defect, while keeping the same simulation environment. The addition of accessories like scissor lifts to adjust the height of the simulator, or illumination and recording systems, whose need emerged during testing with surgeons, is also possible. A synthetic skin on the surface of the rib cage, preventing from looking inside the chest through ribs during the session, and so increasing the overall realism of the simulator, could be easily added, too [9]. Moreover, grippers or in general systems that allow to physically lock the platform to the table and increase the overall stability could be imagined, as suggested by surgeons who test the simulator.

From a testing perspective, only 6 surgeons evaluated the proposed solution, limiting the statistical significance of feedback. Future studies should aim for a larger sample size to obtain more robust data.

References

1. Yoo, S.J., Hussein, N., Barron, D.J.: Congenital heart surgery skill training using simulation models: not an option but a necessity. J. Korean Med. Sci. (2022). https://doi.org/10.3346/jkms.2022.37.e293
2. Barron, D.J., Hussein, N., Yoo, S.J.: Training on Congenital 3D Cardiac Models—Will Models Improve Surgical Performance? (2023). https://doi.org/10.1053/j.pcsu.2022.12.001
3. Bertolini, M., Rossoni, M., Colombo, G.: Operative workflow from ct to 3d printing of the heart: opportunities and challenges. Bioengineering. **8** (2021). https://doi.org/10.3390/bioengineering8100130

4. Bertolini, M., Rossoni, M., Colombo, G.: Additive manufacturing of a compliant multimaterial heart model. Comput. Aided Des. Appl. **19**, 1162–1170 (2022). https://doi.org/10.14733/cadaps.2022.1162-1170
5. Hussein, N., Honjo, O., Haller, C., Hickey, E., Coles, J.G., Williams, W.G., Yoo, S.J.: Hands-on surgical simulation in congenital heart surgery: literature review and future perspective. Semin. Thorac. Cardiovasc. Surg. **32**, 98–105 (2020). https://doi.org/10.1053/j.semtcvs.2019.06.003
6. Peel, B., Lee, W., Hussein, N., Yoo, S.J.: State-of-the-art silicone molded models for simulation of arterial switch operation: innovation with parting-and-assembly strategy. JTCVS Tech. **12**, 132–142 (2022). https://doi.org/10.1016/j.xjtc.2021.12.009
7. Thielen, M., Delbressine, F., Oetomo, S.B., Feijs, L.: Anatomically realistic neonatal heart model for use in neonatal patient simulators. J. Vis. Exp. (2019). https://doi.org/10.3791/56710
8. Mattus, M.S., Ralph, T.B., Pieta Keller, S.M., Waltz, A.L., Bramlet, M.T.: Creation of patient-specific silicone cardiac models with applications in pre-surgical plans and hands-on training. J. Vis. Exp. (2022). https://doi.org/10.3791/62805
9. Peel, B., Voyer-Nguyen, P., Honjo, O., Yoo, S.-J., Hussein, N.: Development of a dynamic chest wall and operating table simulator to enhance congenital heart surgery simulation. 3D Print Med. **6**, 12 (2020). https://doi.org/10.1186/s41205-020-00067-4
10. https://sketchfab.com/3d-models/ventricularseptal-defect-e4e0143741f64025af101658210c8c5d. Last accessed on 24 Jun 2024.
11. https://sketchfab.com/3d-models/human-thorax-4aba9b2ced344bdf8f09656c6a298b50. Last accessed on 24 Jun 2024.

Direct Ink Printing of Liquid Crystalline Elastomers Towards Photoresponsive Artificial Muscles

Greta Feregotto[1], Giovanni Simonetti[2], Ruggero Rossi[1], Daniele Martella[1,2], Caterina Credi[1,3], Camilla Parmeggiani[1,2], and Federico Carpi[4(✉)]

[1] European Laboratory for Non-Linear Spectroscopy (LENS), Florence, Italy
[2] Department of Chemistry, University of Florence, Florence, Italy
[3] National Institute of Optics - National Research Council (INO-CNR), Florence, Italy
[4] Department of Industrial Engineering, University of Florence, Florence, Italy
federico.carpi@unifi.it

Abstract. Smart materials enable the preparation of 3D structures capable of reversible shape changes in response to external stimuli. Among them, liquid crystalline elastomers (LCE) combine the rubber-like mechanical properties of elastomers with the ordered molecular arrangement and sensitivity to external stimuli that are typical of liquid crystals. In this study, a thiol-acrylate Michael addition reaction was used to synthesize a 3D printable nematic photoresponsive LCE ink. Through direct ink writing, LCE strips were 3D printed with a molecular orientation along the direction of printing. The strips displayed tunable photoactuation properties, with ability to generate active stresses up to 13.5 kPa in response to an illumination with blue light at 470 nm and at an optical power density of ~4 mW/mm^2 for 2 s. This makes them usable for applications that require remotely controlled artificial muscles that should operate small loads, such as light-activated soft robots.

Keywords: Liquid Crystalline Elastomer · 3D Printing · Artificial Muscle

1 Introduction

Liquid crystalline elastomers (LCEs) are a class of smart materials that display a combination of properties of both liquid crystals and elastomers. A key feature of LCEs is the orientation of liquid crystal mesogens along the molecular director *n*, which can be altered by an external stimulus, causing macroscopic and anisotropic shape variations [1]. As a result, LCEs are highly elastic and can reversibly deform upon external stimuli, such as variations of temperature [2], light [3] and magnetic field [4]. As such, they are of interest for applications that require soft and lightweight responsive materials for actuation and sensing [5–8].

Among the above mentioned stimuli, light is the most attractive one for applications requiring contactless actuation and a high degree of spatial and temporal resolution. The light-driven response is usually induced by a molecular photoswitch (typically an

© The Author(s), under exclusive license to Springer Nature Switzerland AG 2025
M. Carfagni et al. (Eds.): IDBN 2024, BioSysRob 35, pp. 33–44, 2025.
https://doi.org/10.1007/978-3-031-94626-4_5

azobenzene), incorporated within the polymeric network. In particular, when exposed to light, azobenzenes undergo a change in their spatial arrangement, transitioning from a *trans* to a *cis* configuration [9]. This process causes an increase of entropy at a microscopic level and therefore a decrease in the liquid crystalline order. This mechanism leads to large geometrical deformations, which can be exploited to generate movement or tension [10, 11].

However, to date, the majority of LCE films are fabricated manually, by using liquid crystalline cells [12], highly limiting the final assembly shape, dimensions, and reproducibility, or inducing the alignment trough mechanical stress, which prevent the possibility of creating complex alignments [13].

Therefore, versatile techniques able to make the fabrication of LCE structures simpler, and more reliable are highly desirable. In this sense, 3D printing represents a powerful tool. In particular, in this study we focused our attention on direct ink writing (DIW), as previous studies showed how this technique combined with liquid crystalline elastomers is promising to manufacture devices with customized properties [14, 15]. In DIW, the printing material is extruded through a nozzle, to form a continuous filament, which is then deposited along pre-programmed paths. The shear and elongation forces during the printing process can effectively induce molecular alignments along the printing direction, making it possible to obtain LCE films with pre-defined molecular directors [16, 17].

Nevertheless, to date, most of printed films feature thermal actuation, which can be unsuitable for thermally sensitive working environments (for example in many biomedical applications) and has long response times [18, 19]. On the contrary, photoactivation enables controllable localised activations at higher resolutions and typically faster response times.

Here, we present the development of a 3D printable LCE ink with photoresponsive properties and its use to obtain 3D printed photoactive LCE actuators by DIW.

2 Materials and Methods

2.1 Materials for the Photoresponsive LCE Printable Ink

A photoresponsive LCE ink incorporating azobenzene molecules in the main chain oligomer was developed. The ink consisted of the liquid crystal monomer 1,4-bis-[4-(6-acryloyloxyhexyloxy)benzoyloxy]-2-methylbenzene ("RM82", Whilshire Technologies), the thiol chain extender ethyleneglycolbis-mercaptoacetate ("EGM", Whilshire Technologies), the photoswitch 2-[N-ethyl-4-[(4-nitrophenyl)diazenyl]anilino]ethyl prop-2-enoate ("DR1Acrylate", Sigma-Aldrich Corp.), the photoinitiator 2-Benzyl-2-(dimethylamino)-1-[4-(4-morpholinyl)phenyl]-1-butanone ("Irgacure 369", Sigma-Aldrich Corp.) and the thiol-acrylate Michael addition catalyst triethylamine ("TEA", Sigma-Aldrich Corp.).

The solvent used to dissolve the solid components was dichloromethane ("DCM", Sigma Aldrich Corp.). All the materials were used as received.

2.2 Synthesis of the Photoresponsive LCE Printable Ink

Solid reagents RM82 (1.76 g, 1.1 eq), Irgacure 369 (33.9 mg, 1.5 wt%) and DR1Acrylate (22.6 mg, 1 wt%) were added in a cylindrical vial and dissolved in 1.24 ml of DCM under magnetic stirring at room temperature. The vial containing the mixture was then immediately put into a pre-heated oil bath at 50 °C; then, EGM (381.68 µl, 1.0 eq) and TEA (31.13 µl, 1 wt%) were added dropwise under stirring, to initiate the thiol-acrylate Michael addition reaction. The mixture was then stirred for about 20 min and left in the oil bath for 24 h, to facilitate DCM evaporation. In the end, the ink was dried under low pressure for 30 min or more, until its volume expanded, in order to ensure a complete removal of the solvent.

2.3 Characterization of the Photoresponsive LCE Printable Ink

A preliminary analysis of phase transition temperatures of the ink was performed through polarized optical microscopy, using a cross-polarized microscope (Axio Observer A1, Zeiss) equipped with a hot stage (PE120, Linkam). More accurate measures were obtained using a differential scanning calorimetry (DSC Q-2000, TA Instruments) in a nitrogen atmosphere (2 cycles of cooling and 2 cycles of heating, within the temperature range − 40 °C to 120 °C, at a heating and cooling rate of 20 °C/min). The LC ink phase transition temperatures were evaluated in the first heating cycle at the maximum of endothermic peaks, corresponding to the crystallization temperature T_m and the nematic to isotropic transition temperature T_{NI}, respectively.

2.4 DIW Printing of Photoresponsive LCE Strips

The LCE samples were printed with a 3D printer (Hyrel Engine SR) equipped with a high temperature extrusion head (TAM-15) and a UV LED array. Before printing, the material was loaded into the reservoir and put into an ultrasound bath sonicator at 60 °C for 1 h to remove possible air bubbles, which could affect the quality of the printed samples. The printing head was then loaded into the 3D printer, heated to the printing temperature and equilibrated for about 45 min, until the material started to flow. Luer-lock micronozzles with 0.235 mm inner diameter (26G) were used to print the samples (20x3x0.03 mm rectangular strips).

Three-dimensional models of the printed geometries were created with the software Solidworks, while the G-code file that controlled the printing head movements was generated with the software 3D Slicer.

The LCE strips were printed on a microscope cover glass (24x50x0.1 mm) and were first photopolymerized at a wavelength of 365 nm, using a UV-LED array arranged on the printing head; once the printing procedure was completed, the samples were postcured with a UV light source of 385 nm at 0.2 mW/mm^2 (M385LP1-C4 lamp, Thorlabs) at room temperature for 20 min.

2.5 Characterization of the 3D Printed LCE Strips

The alignment of the printed samples was visually inspected with polarized optical microscopy, by examining variations in birefringence when the orientation of the sample was changed with respect to that of the polarizer.

Transmittance spectra of the ink and the LCE were measured using a PerkinElmer FTIR to compare the characteristic diacrylate signal peaks (811 cm^{-1}, 1410 cm^{-1}, and 1630 cm^{-1}) and to confirm oligomerization through Michael's addition.

Gel fraction tests were carried out as follows. The printed structures were soaked in several solvents (demineralized water, dichloromethane, isopropanol, tetrahydrofuran, toluene, ethanol, ethylacetate, methanol and acetone) for 24 h, then dried at room temperature for 24 h and after that left drying in an oven at 40 °C for 2 h. Subsequently, the average gel fraction was calculated as the ratio between the final and the initial mass three samples for each solvent.

2.6 Thermoactuation of the 3D Printed LCE Strips

Thermoactuation experiments were accomplished using a hot stage (HS82, Mettler Toledo), which allowed for automatically increasing the temperature (at a rate of 10 °C/min) from 25 °C to 220 °C, keeping it constant at each increment for 3 min. The LCE strips were arranged on a microscope cover glass coated with silicone oil and then positioned on the hot plate. The thermoactuation performance of the samples was evaluated by taking images 30 s before the end of each isotherm phase. The images were analyzed to measure the strips' length, using the software ImageJ, by tracking a straight line along the sample's long axis (parallel to the printing direction). The average and standard deviation of LCE thermal contraction was calculated on 5 different samples (dimension 10 x 3 x 0.5 mm).

2.7 Photoactuation of the 3D Printed LCE Strips

The photoresponsive performance of the LCE strips was assessed by measuring both the free stroke (bending) and the blocking force.

For the bending tests, one end of the LCE strip was attached to a sample holder on, whilst the other end was left free to move. One side of the sample was then irradiated (mostly at the level of its central region) with an LED emitting blue light at 470 nm (M470L5 lamp, Thorlabs) and at an intensity of 5.1 mW/mm^2. The sample was exposed to multiple cycles of irradiation, during which the LED was switched on for 3, 6, 9, 12 and 15 s, and off for 30 s in between. The absolute and relative bending angle were determined by analyzing video frames with the software Greenshot and ImageJ.

The blocking force developed by the strips was measured using a custom-made setup, consisting of a force transducer (KG4A, Scientific Instruments, Germany) and two LEDs emitting blue light at 470 nm (M470L5 lamp, Thorlabs). The sample was fixed to the force transducer and to a sample holder. Two LEDs were used to irradiate the sample simultaneously on both sides. A customized LabVIEW program controlled the LEDs light intensity and their activation time and recorded the force signal. The measured force was normalized by the strip cross sectional area to calculate the developed tension.

3 Results and Discussion

3.1 Photoresponsive LCE Printable Ink

The LCE mixture (Fig. 1A) was synthesized via the thiol-acrylate Michael addition reaction with a slight excess of acrylate as compared to thiol. Triethylamine (TEA) served as a catalyst to facilitate the oligomerization process (Fig. 1B). The constitutive materials and the synthesis are detailed in Materials and Methods.

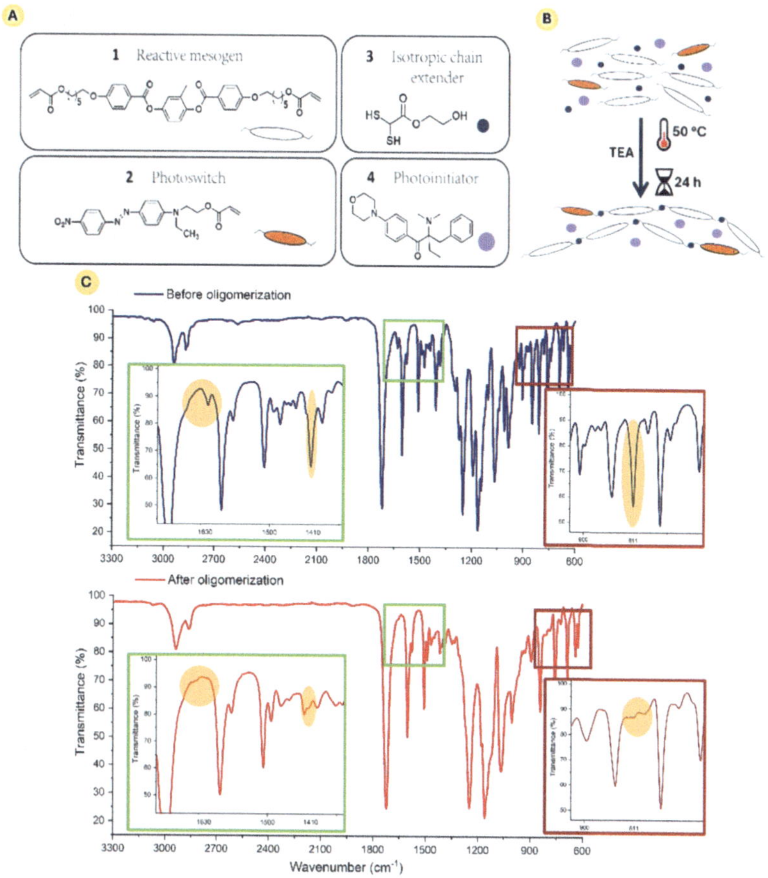

Fig. 1. Photoresponsive LCE printable ink: A) Chemical composition of the molecules used; B) Schematic representation of the oligomerization reaction; C) ATR spectra of the ink before and after oligomerization; after the completion of the reaction, the acrylate peaks (811 cm^{-1}, 1410 cm^{-1} and 1630 cm^{-1}) decreased, as a consequence of the development of the polymeric chain.

Attenuated total reflection (ATR) spectra of the mixture before the addition of the catalyst and after 24 hours (Fig. 1C) revealed that the reaction occurred successfully. Indeed, after chain extension, the signals corresponding to the acrylate peaks (811 cm^{-1},

1410 cm^{-1} and 1630 cm^{-1}), which were visible in the spectra before oligomerization, decreased, implying the development of the polymeric chain.

The oligomer was later crosslinked by light-initiated free radical photocuring, by using a photoinitiator, as detailed in Materials and Methods.

Differential scanning calorimetry traces showed a nematic to isotropic phase transition temperature (T_{NI}), both in heating and cooling, around 76 °C (Fig. 2A).

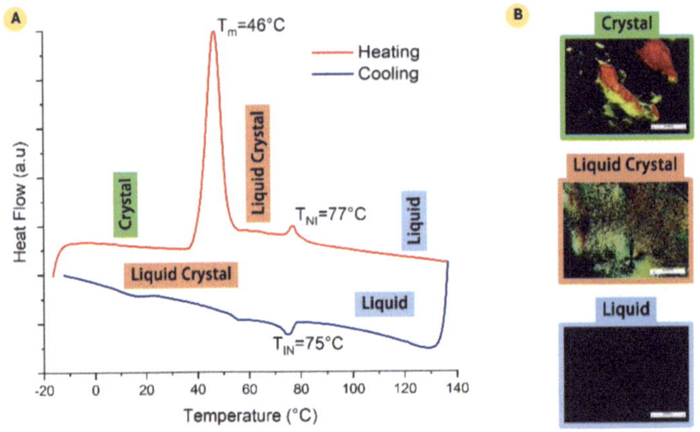

Fig. 2. Phase transitions temperatures of the ink: A) Differential scanning calorimetry trace of the LCE mixture; B) Polarized optical microscopy images of the different phase textures.

The ink's textures at different temperatures were visually inspected through polarized optical microscopy, as shown with representative images in Fig. 2B. The oligomer is crystalline at room temperature (upper panel), while above 46 °C a nematic phase (middle panel) is formed up to 77 °C, when an isotropic phase can be observed (bottom panel). Interestingly, on cooling, crystallization does not occur for several hours, allowing for the simple use of the ink also at room temperature for long writing processes.

3.2 Direct Ink Writing of Photoresponsive LCE Strips

The ink was used for DIW to 3D print monodomain-aligned LCE strips, as detailed in Materials and Methods.

The molecular alignment was determined by shear stresses arising within the printing tip during extrusion. Accordingly, the mesogens were oriented along the printing direction, resulting in a uniaxially aligned LCE strip.

The printing parameters were empirically tuned according to the quality of the prints. In particular, two parameters proved critical for producing uniform and aligned samples: 1) printing velocity, which was correlated with the shear stresses and therefore with the molecular orientation; 2) extrusion temperature, which was essential to maintain the material in its nematic phase.

The printing velocity was kept within the range of 4–5 mm/s, as lower values led to non-aligned structures, while higher values did not let the material flow properly, leaving empty spaces in the final structure. The extrusion temperature was kept between 60 °C and 65 °C, to set the nozzle temperature 10–15 °C below the nematic-to-isotropic phase transition temperature (77 °C for our materials), as suggested by Del Pozo et al. [20], preserving the nematic phase and guaranteeing a sufficiently low viscosity of the ink during the printing. The printing bed was maintained at 35 °C.

With those parameters, it was possible to print LCE strips (20x3x0.03 mm) on a microscope cover glass. During printing, the molecular alignment along the print path was fixed by a first photocuring, induced by a UV-LED array arranged on the printing head. After printing, the strips were permanently crosslinked under a UV lamp at room temperature. At the end of the process, the correct molecular alignment within the printed strips was verified using polarized optical microscopy. Then, the strips were detached from the glass substrate (using a wet blade), to obtain free-standing samples. The whole fabrication process is schematically represented in Fig. 3.

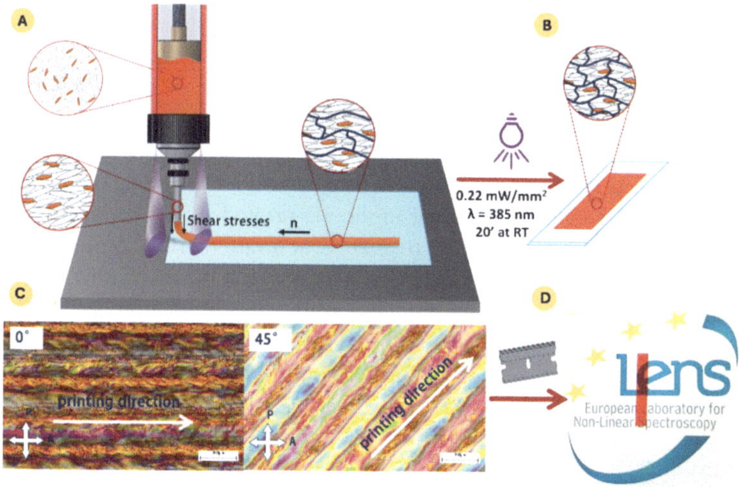

Fig. 3. Overview of the fabrication process to obtain 3D printed LCE strips: A) DIW of the ink on top of a microscope cover glass; B) Postcuring of the printed strip; C) Visual inspection of the surface alignment with polarized optical microscopy; D) Detachment of the printed sample from the glass (the inset photo shows the transparency of a sample).

To verify the successful completion of the photocuring, the gel fraction (percentage of polymer involved in the crosslinked network) was determined, as detailed in Materials and Methods. As illustrated in Fig. 4, the gel fraction was above 95% for all the tests performed, except for dichloromethane, which was still resulting in a gel fraction above 80%, thereby indicating that the polymeric chains were properly crosslinked.

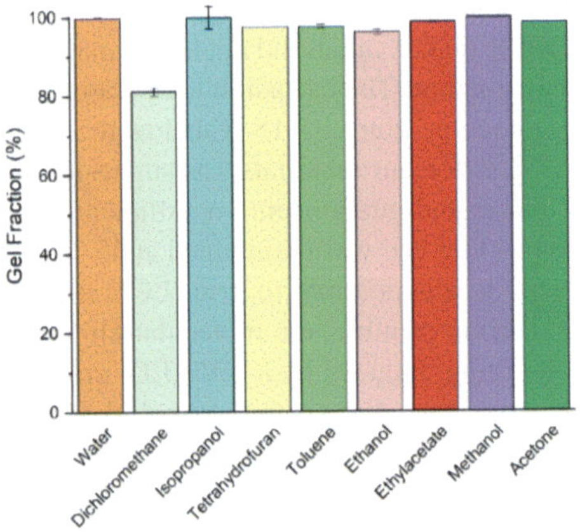

Fig. 4. Gel fraction of the 3D printed LCE strips. Th the error bars represent the standard deviation among 3 samples.

3.3 Actuation of the 3D Printed LCE Strips

Thermal Actuation

In general, LCE typically undergo structural changes when heated above the T_{NI}. This behavior is related to a decrease of the orientational order degree of the mesogens, leading to a contraction along the molecular director *n* and an expansion along the perpendicular directions. To investigate such behavior in the 3D printed strips, samples were heated on a hot plate from room temperature to 220 °C at a rate of 10 °C/min. The resulting temperature-dependent contraction in length (along which the molecules were aligned) is plotted in Fig. 5A. At 220 °C, the samples exhibited a contraction of 10% and when they were cooled down to room temperature, they recovered 95% of their initial length (Fig. 5B). This structural reversibility further confirms that the polymeric network is well developed.

Photoactuation

As recalled in the Introduction, LCE containing photoresponsive molecules, like azobenzene, are sensitive to photostimulations, such that structural changes can be induced, in this case, by light. This phenomenon is due to a *trans-cis* isomerization of the dye, which increases the mesogenic disorder, causing a macroscopic deformation. To investigate such a behavior in the 3D printed LCE strips, both free stroke and blocking force tests were performed.

The former was accomplished as follows. The strips (10 × 3 × 0.03 mm) were hung on a sample holder, constraining the higher end and keeping the lower end free to move, as shown in Fig. 6A.

Illumination of one side of the sample with a blue light at 470 nm and a power density of 5.1 mW/mm^2 caused bending of the strip towards the light (Fig. 6B). This behavior was the consequence of a light intensity gradient throughout the sample thickness [21]. Once the stimulus was switched off, the LCE strip did not completely return to the initial configuration, retaining some residual strain (Fig. 6B).

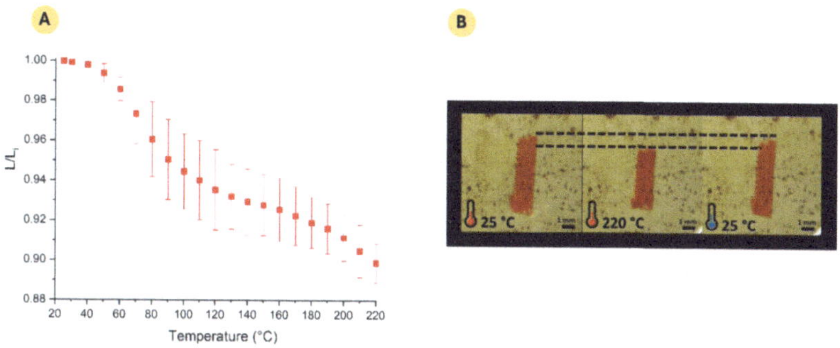

Fig. 5. Thermoactuation of the 3D printed LCE strips: A) Temperature-induced contraction of the samples along their length, expressed as the ratio between the instantaneous length L and its initial value L_i; the error bars represent the standard deviation among 5 samples; B) Example of thermally induced contraction of a sample, which was heated from 25 °C to 220 °C and then cooled back to 25 °C.

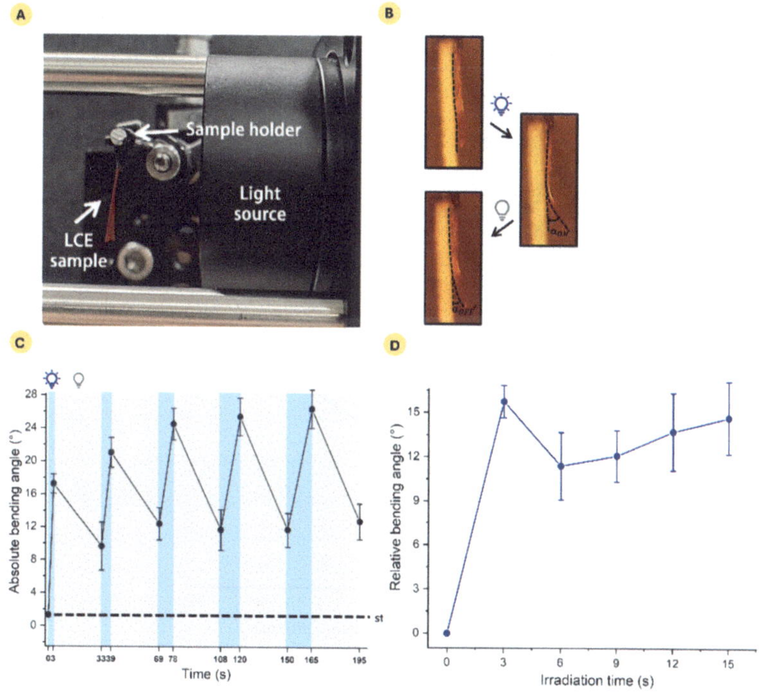

Fig. 6. Photoactivated free stroke of the 3D printed LCE strips: A) Test setup; B) Example of bending of the strip in response to a photostimulation; C) variation of the absolute bending angle for increasing durations of an irradiation with blue light at 470 nm and at a power density of 5.1 mW/mm^2; D) Dependence of the relative bending angle on the irradiation time. The error bars represent the standard deviation among 5 samples.

By optically stimulating the sample repeatedly and with increasing durations, the absolute bending angle tended to increase, although showing a saturation effect (Fig. 6B). Figure 6D presents the dependence on the irradiation time of the relative bending angle, defined as the difference between the angle achieved at the end of each illumination phase (light on) and the angle achieved at the end of the preceding relaxation phase (light off).

It is worth noting that the power density used in these tests (5.1 mW/mm^2) was the highest that could be sustained by samples. Greater power densities were found to lead to structural degradations of the material.

The blocking force tests were performed as follows. The strips (20 × 1.5 × 0.03 mm) were tested under isometric conditions using a custom-made setup, shown in Fig. 7A.

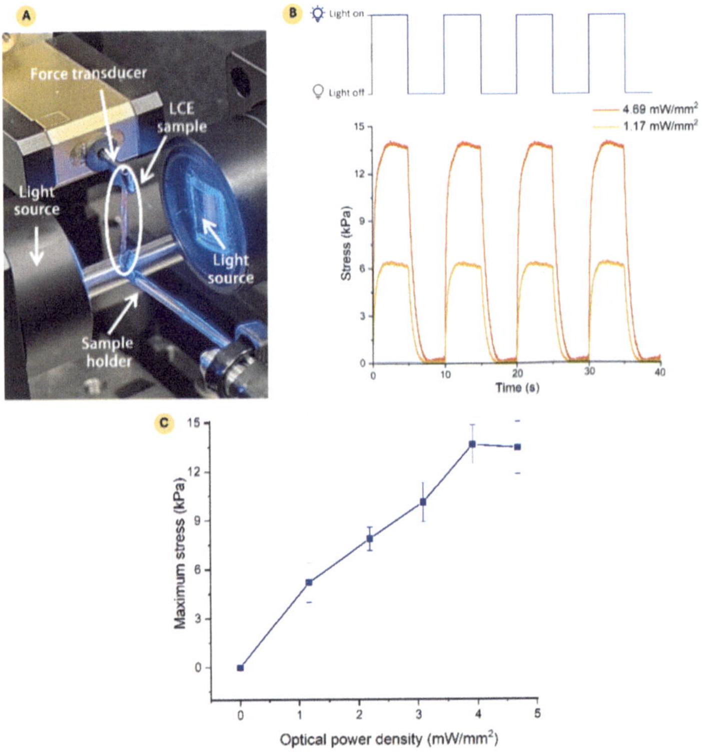

Fig. 7. Photoactivated blocking stress of the 3D printed LCE strips: A) Test setup; B) Examples of stress signals in response to cyclic illuminations at different power densities; C) Dependence of the maximum stress on the optical power density; the error bars represent the standard deviation among 5 samples.

The samples were arranged vertically, attaching the top end to a force transducer and the bottom end to a sample holder. Both sides of the strip were illuminated simultaneously using two lateral LEDs, which emitted light at a wavelength of 470 nm, and the force exerted by the sample along its length (direction of molecular alignment) was recorded. Specifically, the stimulations were done with pulsed light: during a cycle of measurements, the LEDs were on for 5 s and off for 5 s. Fig. 7B presents examples of stress signals gathered in response to illuminations with different power densities. As presented in Fig. 7C, the maximum recorded stress showed a monotonic increase for increasing optical power densities, as expected [22]. The highest value of active stress generated by the strips was 13.5 kPa in response to illumination at ~4 mW/mm^2 for 2 s (Fig. 7C).

4 Conclusions

We presented the synthesis of a 3D printable nematic photoresponsive LCE ink and the demonstration of 3D printing of photoactive LCE strips by DIW. The 3D fabrication process allowed for programming the molecular order along the direction of the printing path. The printed samples exhibited not only thermo- but also photo-responsive actuation properties, with ability to generate active stresses up to 13.5 kPa at ~4 mW/mm^2. This makes them usable for applications that require remotely controlled artificial muscles that should operate small loads, such as light-activated soft robots.

Future studies should aim at improving the material's actuation properties, for example by introducing within the LCE mixture functional particles.

Acknowledgments. The research leading to these results has received funding from the European Union's Horizon 2020 research and innovation program under grant agreement No 952166 (REPAIR).

References

1. Terentjev, E.M.: Liquid-crystalline elastomers. J. Phys. Condens. Matter. **11**(24), R239–R257 (1999)
2. White, T.J., Broer, D.J.: Programmable and adaptive mechanics with liquid crystal polymer networks and elastomers. Nat. Mater. **14**(11), 1087–1098 (2015)
3. Ikeda, T., Mamiya, J.I., Yu, Y.: Photomechanics of liquid-crystalline elastomers and other polymers. Angew. Chem. Int. Ed. **46**(4), 506–528 (2007)
4. Winkler, M., Kaiser, A., Krause, S., Finkelmann, H., Schmidt, A.M.: Liquid crystal elastomers with magnetic actuation. In: Macromolecular Symposia, vol. 291, No. 1, pp. 186–192. WILEY-VCH Verlag, Weinheim (2010)
5. Li, S., Bai, H., Liu, Z., Zhang, X., Huang, C., Wiesner, L.W., et al.: Digital light processing of liquid crystal elastomers for self-sensing artificial muscles. Sci. Adv. **7**(30), eabg3677 (2021)
6. Ma, J., Yang, Y., Zhang, X., Xue, P., Valenzuela, C., Liu, Y., et al.: Mechanochromic and ionic conductive cholesteric liquid crystal elastomers for biomechanical monitoring and human–machine interaction. Mater. Horiz. (2024)
7. Han, W.C., Lee, Y.J., Kim, S.U., Lee, H.J., Kim, Y.S., Kim, D.S.: Versatile mechanochromic sensor based on highly stretchable chiral liquid crystalline elastomer. Small. **19**(7), 2206299 (2023)
8. Lyu, P., Astam, M.O., Sánchez-Somolinos, C., Liu, D.: Robotic pick-and-place operations in multifunctional liquid crystal elastomers. Adv. Intell. Syst. **4**(12), 2200280 (2022)

9. Bandara, H.D., Burdette, S.C.: Photoisomerization in different classes of azobenzene. Chem. Soc. Rev. **41**(5), 1809–1825 (2012)
10. Ferrantini, C., Pioner, J.M., Martella, D., Coppini, R., Piroddi, N., Paoli, P., et al.: Development of light-responsive liquid crystalline elastomers to assist cardiac contraction. Circ. Res. **124**(8), e44–e54 (2019)
11. Nocentini, S., Parmeggiani, C., Martella, D., Wiersma, D.S.: Optically driven soft micro robotics. Adv. Opt. Mater. **6**(14), 1800207 (2018)
12. Donato, S., Martella, D., Salzano de Luna, M., Arecchi, G., Querceto, S., Ferrantini, C., et al.: The role of crosslinker molecular structure on mechanical and light-actuation properties in liquid crystalline networks. Macromol. Rapid Commun. **44**(9), 2200958 (2023)
13. Yakacki, C.M., Saed, M., Nair, D.P., Gong, T., Reed, S.M., Bowman, C.N.: Tailorable and programmable liquid-crystalline elastomers using a two-stage thiol–acrylate reaction. RSC Adv. **5**(25), 18997–19001 (2015)
14. Guan, Z., Wang, L., Bae, J.: Advances in 4D printing of liquid crystalline elastomers: materials, techniques, and applications. Mater. Horiz. **9**(7), 1825–1849 (2022)
15. Wang, Z., Wang, Z., Zheng, Y., He, Q., Wang, Y., Cai, S.: Three-dimensional printing of functionally graded liquid crystal elastomer. Sci. Adv. **6**(39), eabc0034 (2020)
16. Chen, M., Gao, M., Bai, L., Zheng, H., Qi, H.J., Zhou, K.: Recent advances in 4D printing of liquid crystal elastomers. Adv. Mater. **2209566** (2023)
17. Wang, Z., Guo, Y., Cai, S., Yang, J.: Three-dimensional printing of liquid crystal elastomers and their applications. ACS Appl. Polym. Mater. **4**(5), 3153–3168 (2022)
18. Li, S., Song, Z., Fan, Y., Wei, D., Liu, Y.: Four-dimensional printing of temperature-responsive liquid crystal elastomers with programmable shape-changing behavior. Biomimetics. **8**(2), 196 (2023)
19. Saed, M.O., Ambulo, C.P., Kim, H., De, R., Raval, V., Searles, K., et al.: Molecularly-engineered, 4D-printed liquid crystal elastomer actuators. Adv. Funct. Mater. **29**(3), 1806412 (2019)
20. Del Pozo, M., Sol, J.A., Schenning, A.P., Debije, M.G.: 4D printing of liquid crystals: what's right for me? Adv. Mater. **34**(3), 2104390 (2022)
21. Kondo, M., Yu, Y., Ikeda, T.: How does the initial alignment of mesogens affect the photoinduced bending behavior of liquid-crystalline elastomers? Angew. Chem. Int. Ed. **45**(9), 1378–1382 (2006)
22. Vitale, G., Grandinetti, B., Querceto, S., Martella, D., Tesi, C., Poggesi, C., et al.: Photoresponsive polymer-based biomimetic contractile units as building block for artificial muscles. Macromol. Mater. Eng. **307**(9), 2200187 (2022)

Surface Characteristics Improvement of 3D Printed 316 L Steel for Biomedical Applications

Eleonora Bolli[1], Saulius Kaciulis[1], Alex Lanzutti[2], Alessio Mezzi[1], Roberto Montanari[3], Alessandra Palombi[3](✉), Francesco Sordetti[2], Emanuele Vaglio[2], Alessandra Varone[3], Claudio Verona[3], and Ali Zahid[3]

[1] ISMN-CNR, P.O. Box 10, Monterotondo Stazione, 00015 Rome, Italy
[2] Polytechnic Department of Engineering and Architecture, University of Udine, Via delle Scienze 208, 33100 Udine, Italy
[3] Department of Industrial Engineering, University of Rome "Tor Vergata", Via del Politecnico 1, 00133 Rome, Italy
alessandra.palombi@uniroma2.it

Abstract. Owing to its high corrosion resistance 316 L steel is widely used in biomedical implants, such as screws and plates. To improve wear resistance in physiological environment, surface treatments are often carried out, and among them low temperature carburizing is one of the most effective.

This work optimized a plasma assisted low temperature carburizing treatment made by using a 2.5% CO_2 + 97.5% H_2 gas mixture. The scope was to promote the formation of the S-phase (expanded austenite) avoiding the formation of a diamond like carbon (DLC) over-layer. Treatments with different duration (30, 45, 60, 90, and 120 minutes) have been investigated with significant hardness increase in all the samples as compared to original material. After 60 minutes the DLC over-layer starts to grow thus this treatment time seems to be the best trade-off for hardening the surface without the formation of DLC that, being hard and brittle, may give rise to debris with detrimental effects on wear resistance.

Keywords: 316 L steel · Additive manufacturing · Plasma assisted low temperature carburizing · Surface characteristics

1 Introduction

In biomedical applications, such as orthopedic implants, the biocompatible material must exhibit good wear and corrosion resistance in contact with body fluids [1, 2].

Among metallic materials Ti6Al4V, CoCrMo, and AISI 316 L stainless steel are the most used for the production of implants and medical devices [2, 3]. 316 L implants are cheaper than those made of Ti and Co-based alloys [1] but this steel presents poor wear behavior and is prone to localized corrosion attack in physiological conditions [4]. This can lead to the release of debris and metal ions, harmful for the human body [5].

The application of Additive Manufacturing (AM) processes, such as Laser Powder Bed Fusion (L-PBF), is gaining increasing interest for the production of components

© The Author(s), under exclusive license to Springer Nature Switzerland AG 2025
M. Carfagni et al. (Eds.): IDBN 2024, BioSysRob 35, pp. 45–52, 2025.
https://doi.org/10.1007/978-3-031-94626-4_6

in many different fields [6–10]. In biomedicine AM offers the possibility to produce implants customized to fit the patient anatomy.

The microstructure of metallic alloys produced through AM is different from those obtained by conventional techniques, resulting in higher mechanical properties [11–16], especially with an optimization of the printing parameters [17, 18] and the post process heat treatments [19]. Nonetheless, wear behavior still represents a critical issue for austenitic stainless steels and thermo-chemical treatments are necessary to improve surface hardness.

Conventional treatments, such as carburizing, are not suitable for austenitic stainless steels because the temperature (above 550 °C), leads to the precipitation of Cr carbides with a consequent detrimental effect on corrosion resistance. Therefore, low-temperature treatments have been developed to avoid this drawback [20], however they involve long treatment times. Exploiting plasma properties, present authors developed an efficient carburizing treatment leading to improved surface hardness with shorter treatment time. The treatment leads to excellent results in both conventionally [21–24] and additively manufactured steel [25–27]. This treatment, carried out for 7 hours with an atmosphere containing 2.5% (CH_4) + 97.5% (H_2), induces the formation of a ~ 25 μm thick layer of expanded austenite (S-phase), responsible of the surface hardening and corrosion resistance improvement, covered by a ~ 2 μm thick Diamond-Like Carbon (DLC) over-layer. The hard and brittle DLC over-layer partly improves the surface hardness but its tendency to detach from the surface during tribological tests is detrimental to wear resistance [26].

Since the growth of DLC starts when the metal lattice has been completely saturated by C, this work investigated new treatments of shorter time (30, 45, 60, 90 and 120 minutes) made by using the same gas mixture. The scope was to enhance hardness by forming S-phase without the DLC over-layer. The samples were examined by X-ray diffraction (XRD), Atomic Force Microscopy (AFM) observations and micro-hardness tests.

2 Materials and Methods

2.1 Sample Preparation

316 L stainless steel (C 0.024, Cr 16.87, Mo 2.06, N 0.083, Mn 1.35, Si 0.40, Ni 10.05, P 0.031, S 0.029, Fe to balance wt.%) samples were produced by using Laser Powder Bed Fusion (L-PBF) technique.

Blocks (30.5 × 20.5 × 7 mm^3) were 3D printed, in an inert atmosphere of Ar, on a Concept Laser M2 Cusing machine (laser power P = 180 W, scanning speed v_s = 600 mm/s, laser spot diameter d_l = 120 μm, hatch distance h_d = 105 μm, layer thickness t_l = 25 μm). Details of printing strategy are reported in ref. [26]. The samples (7 × 6 × 2 mm^3) were cut from the blocks, then a mirror-like surface was obtained by mechanical polishing; before plasma treatment, each sample was immersed for 10 minutes in a sonication bath.

The microstructure of the printed material consisted of austenite and a residual δ-ferrite (~5% maximum content).

2.2 Plasma Treatments

The samples were plasma treated with a 2.5% CH_4 plus 97.5% H_2 gas mixture for 30, 45, 60, 90 and 120 minutes. Surface treatments were performed using a microwave plasma-enhanced chemical vapor deposition (CVD) reactor [28]. Before each plasma treatment, the surface was bombarded with H_2 for 15 minutes to remove the passive oxidation layer. To obtain a temperature ~ 475 °C, the chamber pressure and microwave power have been kept at a constant value of approximately 70 mbar and in the range 400–500 W, respectively. This temperature has been proved to be a good trade-off to reach good surface hardness and to avoid the precipitation of $M_{23}C_6$ carbides and other undesired phases (η, χ and σ) [25].

2.3 Sample Characterization

XRD measurements were performed by using a PW 1729 diffractometer (Philips, Eindhoven, The Netherlands) with Mo-Kα radiation (λ = 0.07093 nm), 2θ steps of 0.005° and counting time of 10 s per step. From precision peak profiles the C content C_γ (wt%) in the S-phase has been determined through the empirical relationship proposed by Ridley et al. [29]:

$$a_\gamma = a_{0\gamma} + \alpha C_\gamma \tag{1}$$

being $a_{0\gamma}$ and a_γ the lattice parameters of the untreated austenite and S-phase, respectively, and α a constant (0.0044 nm/wt% C).

Atomic Force Microscopy (FlexAFM, Nanosurf Liestal, Switzerland) operating in dynamic force mode has been employed to investigate the surface morphology of the samples.

Vickers micro-hardness tests (micro-hardness tester Shimadzu Corporation, Kyoto, Japan) have been carried out to determine the hardness vs. penetration depth profiles by using increasing loads (25, 50, 100, 200, 300, 500 and 1000 g).

3 Results and Discussion

3.1 XRD Analysis

XRD analysis has been used to investigate the formation of the S-phase layer, related to C enrichment leading to a lattice expansion with consequent hardness enhancement. Table 1 shows the peaks positions and the calculated C content of all the examined samples. The lattice expansion gives rise to the shift of γ_{111} and γ_{200} XRD peaks towards lower angles and from eq. 1 the C content in the matrix has been calculated with an error of ±0.1 wt.%. Not Treated (NT) sample has been used as a reference to determine $a_{0\gamma}$.

In the samples treated for 30 minutes, the average C content in a layer of ~25 μm from the surface (that analyzed by XRD and corresponding to the S-phase thickness) is ~2.2 wt.%, namely much higher than the nominal value in the as-built material, indicating that S-phase forms even after the shortest treatment time examined here. C content increases with treatment time up to 90 minutes, then it remains nearly constant (within the experimental error) because austenite is not able to host further C in interstitial positions.

Table 1. Peak position, lattice parameter and C content of each sample.

	2θ (deg)		a (nm)	C wt.%
	γ_{111}	γ_{200}		
Not treated	19.90	22.97	0.355	–
30 minutes	19.33	22.25	0.366	2.19
45 minutes	19.45	22.37	0.364	2.17
60 minutes	19.38	22.11	0.365	2.30
90 minutes	19.48	22.21	0.363	2.42
120 minutes	19.48	22.21	0.363	2.42
420 minutes	19.35	22	0.366	2.32

3.2 AFM Observations

Figure 1 shows the surface morphology of the non-treated sample, and of samples treated for 60, 90 and 120 minutes.

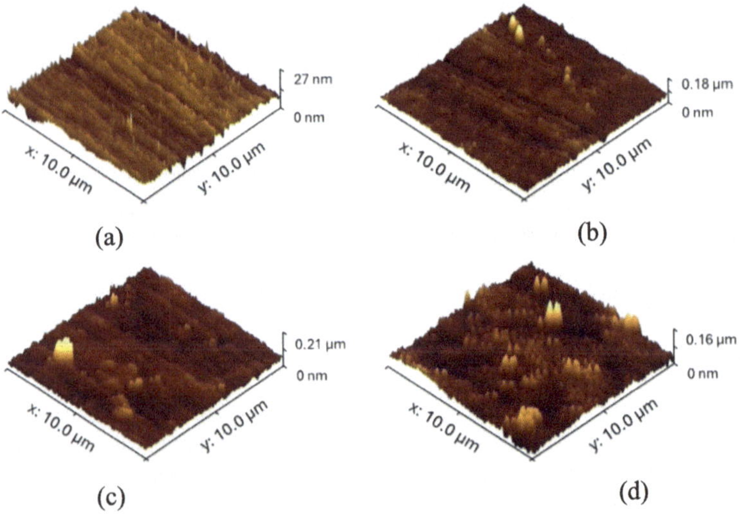

Fig. 1. AFM images of the non-treated sample (a), and of samples plasma treated for 60 (b), 90 (c), and 120 (d) minutes.

The surface of the non-treated sample (Fig. 1a) presents only some shallow scratches, typical features due to the mechanical polishing process. Following plasma treatments the surface morphology changes depending on the treatment time. After 60 (Fig. 1v) and 90 (Fig. 1c) minutes, the marks of the polishing are still visible, the surface appears nearly etched from the preliminary H_2 bombardment in the CVD reactor, some small spot of DLC can be observed but the over-layer is not fully structured. The complex DLC morphology of peaks and valleys [26] can be clearly recognized only on the surface of samples treated for 120 minutes (Fig. 1d) or more.

3.3 Micro-hardness Tests

Micro-hardness tests were carried out on the treated surface by employing increasing loads to measure the hardness profile near the surface (Fig. 2). For a homogeneous material the Vickers micro-hardness value does not depend on the load employed in the test. In the plasma treated samples the C content decreases from the surface towards the inner part of the material and consequently micro-hardness decreases too. The S-phase layer and the underlying unaffected austenite give a different contribution to the average value with increasing penetration depth: for lower penetration depths, i.e. lower loads, the harder S-phase layer gives the principal contribution.

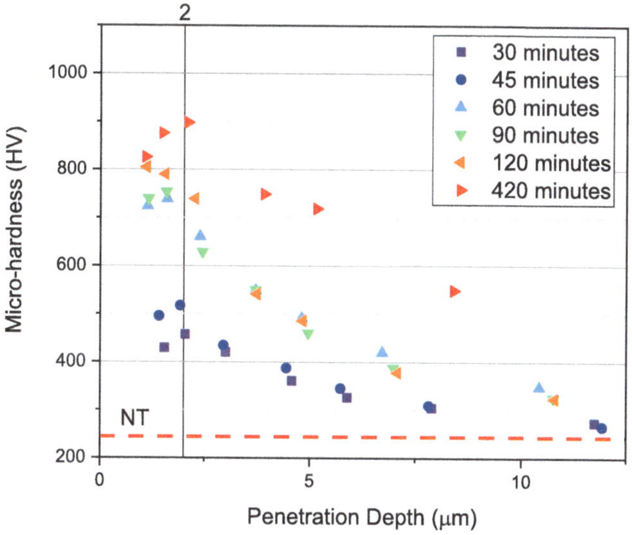

Fig. 2. Vickers micro-hardness vs penetration depth of treated samples. Data of 7 hours of treatment, taken from a previous work [25], and of NT sample are reported for comparison.

Moreover, DLC has an intrinsic hardness (~ 900 HV) higher than S-phase, therefore micro-hardness tests that involve penetration depths <2 μm give information on the deposition of this over-layer. Figure 2 shows that the hardness in this very thin layer strongly increases with the treatment time passing from ~410 HV (45 minutes) to ~900 HV (420 minutes) and confirms the AFM observations.

Comparing the Vickers micro-hardness values obtained by using smaller loads (penetration depths <2 μm) samples treated for 120 and 420 minutes exhibit higher micro-hardness values, greater than 800 HV, due to the formation of the undesired DLC over-layer.

Increasing the treatment time from 60 to 90 to 120 minutes does not lead to changes in S-phase properties, only promotes the deposition of a DLC over-layer. Between 90 and 120 minutes the C content in the S-phase reaches a saturation limit, no more C can enter the matrix and, therefore, it starts to deposit on the surface. It has already been observed that longer treatment times, for example for 7 hours, can further increase surface hardness, but the formation of the DLC over-layer cannot be avoided. Therefore, it is concluded that a treatment time of 60 minutes is the optimal one because it guarantees the formation of S-phase without the DLC over-layer.

4 Conclusions

Low-temperature plasma-assisted treatments in 2.5% (CH4) + 97.5%(H2) atmosphere have been carried out for 30, 45, 60, 90 and 120 minutes on additively manufactured 316 L stainless steel, largely used in the production of orthopedic implants.

- Expanded austenite layer already forms after 30 minutes of treatment.
- C content in the S-phase reaches a maximum value after 60 minutes, then it remains nearly constant.
- The results obtained for treatment time longer than 60 minutes indicate that the main effect is the formation of the undesired DLC without significant changes of the S-phase characteristics.
- On these grounds the treatment of 60 minutes is believed to be the optimal one. Wear and corrosion tests are underway to assess the conclusions.

References

1. Ali, S., Irfan, M., Niazi, U.M., Rani, A.M.A., Rashedi, A., Rahman, S., Khan, M.K.A., Alsaiari, M.A., Legutko, S., Petrů, J., Trefil, A.: Microstructure and mechanical properties of modified 316L stainless steel alloy for biomedical applications using powder metallurgy. Materials. **15**(8), 2822 (2022)
2. Bhat, S.V.: Overview of biomaterials. In: Bhat, S.V. (a c. di) (ed.) Biomaterials, pp. 1–11. Springer, Dordrecht (2002)
3. Mahapatro, A.: Metals for biomedical applications and devices. J. Biomater. Tissue Eng. **2**, 259–268 (2012)
4. Asri, R.I.M., Harun, W.S.W., Samykano, M., Lah, N.A.C., Ghani, S.A.C., Tarlochan, F., Raza, M.R.: Corrosion and surface modification on biocompatible metals: a review. Mater. Sci. Eng. C. **77**, 1261–1274 (2017)
5. Tracana, R.B., Sousa, J., Carvalho, G.S.: Mouse inflammatory response to stainless steel corrosion products. J. Mater. Sci. Mater. Electron. **5**, 596–600 (1994)
6. Moridi, A.: Biomedical applications of metal additive manufacturing: current state-of-the-art and future perspective. Am. J. Biomed. Sci. Res. **7**(1), 6–10 (2020)
7. Singh, S., Ramakrishna, S.: Biomedical applications of additive manufacturing: present and future. Curr. Opin. Biomed. Eng. **2**, 105–115 (2017)
8. Vignesh, M., Ranjith Kumar, G., Sathishkumar, M., Manikandan, M., Rajyalakshmi, G., Ramanujam, R., Arivazhagan, N.: Development of biomedical implants through additive manufacturing: a review. J. Mater. Eng. Perform. **30**, 4735–4744 (2021)

9. Montanari, R., Palombi, A., Richetta, M., Varone, A.: Additive manufacturing of light alloys for aerospace: an overview. In: Ceccarelli, M., Santo, L., Paoloni, M., e Cupertino, G. (eds.) Design advances in aerospace robotics, pp. 110–128. Springer, Cham (2023)
10. Montanari, R., Palombi, A., Richetta, M., Varone, A.: Additive manufacturing of aluminum alloys for aeronautic applications: advantages and problems. Metals. **13**(4) (2023)
11. Tocci, M., Varone, A., Montanari, R., Pola, A.: Study of high temperature properties of AlSi10Mg alloy produced by laser-based powder bed fusion. Mater. Sci. Forum. **1016**, 1485–1491 (2021)
12. Cabibbo, M., Montanari, R., Pola, A., Tocci, M., Varone, A.: Mechanical spectroscopy study of as-cast and additive manufactured AlSi10Mg. J. Alloys Compd. **914**, 165361 (2022)
13. de Crescenzo, C., Fava, A., Montanari, R., Palombi, A., Varone, A.: Mechanical spectroscopy of biocompatible metal alloys. In: Montanari, R., Richetta, M., Febbi, M., Staderini, E.M. (eds.) Engineering Methodologies for Medicine and Sports. EMMS 2024. Mechanisms and Machine Science, 162. Springer, Cham (2024)
14. Choo, H., Sham, K.L., Bohling, J., Ngo, A., Xiao, X., Ren, Y., Depond, P.J., Matthews, M.J., Garlea, E.: Effect of laser power on defect, texture, and microstructure of a laser powder bed fusion processed 316L stainless steel. Mater. Des. **164**, 107534 (2019)
15. Riabov, D., Leicht, A., Ahlström, J., Hryha, E.: Investigation of the strengthening mechanism in 316L stainless steel produced with laser powder bed fusion. Mater. Sci. Eng. A. **822**, 141699 (2021)
16. Bartolomeu, F., Buciumeanu, M., Pinto, E., Alves, N., Carvalho, O., Silva, F.S., Miranda, G.: 316L stainless steel mechanical and tribological behavior—a comparison between selective laser melting, hot pressing and conventional casting. Addit. Manuf. **16**, 81–89 (2017)
17. Lanzutti, A., Sordetti, F., Montanari, R., Varone, A., Marin, E., Andreatta, F., Maschio, S., Furlani, E., Magnan, M., Vaglio, E., Pakhomova, E., Sortino, M., Totis, G., Fedrizzi, L.: Effect of powder recycling on inclusion content and distribution in AISI 316L produced by L-PBF technique. J. Mater. Res. Technol. **23**, 3638–3650 (2023)
18. Ahmed, N., Barsoum, I., Haidemenopoulos, G., Al-Rub, R.K.A.: Process parameter selection and optimization of laser powder bed fusion for 316L stainless steel: a review. J. Manuf. Process. **75**, 415–434 (2022)
19. Lemke, J.N., Casati, R., Lecis, N., Andrianopoli, C., Varone, A., Montanari, R., Vedani, M.: Design of wear-resistant austenitic steels for selective laser melting. Metall. Mater. Trans. A Phys. Metall. Mater. Sci. **49**, 962–971 (2018)
20. Collins, S.R., Williams, P.C.: Low-temperature colossal supersaturation. Adv. Mater. Process. **164**, 32–33 (2006)
21. Ciancaglioni, I., Donnini, R., Kaciulis, S., Mezzi, A., Montanari, R., Ucciardello, N., Verona-Rinati, G.: Surface modification of austenitic steels by low temperature carburization. Surf. Interface Anal. **44**, 1001–1004 (2012)
22. Balijepalli, S.K., Donnini, R., Kaciulis, S., Montanari, R., Varone, A.: Young's modulus profile in Kolsterized AISI 316L steel. Mater. Sci. Forum. **762**, 183–188 (2013)
23. Balijepalli, S.K., Colantoni, I., Donnini, R., Kaciulis, S., Lucci, M., Montanari, R., Ucciardello, N., Varone, A.: Elastic modulus of S phase in kolsterized 316L stainless steel. Metallurgia Italiana. **105**, 42–47 (2013)
24. Balijepalli, S.K., Ceschini, L., Ciancaglioni, I., Kaciulis, S., Mezzi, A., Montanari, R., Martini, C., Verona Rinati, G.: Corrosion effect to the surface of stainless steel treated by two processes of low temperature carburization. Surf. Interface Anal. **46**, 731–734 (2014)
25. Montanari, R., Lanzutti, A., Richetta, M., Tursunbaev, J., Vaglio, E., Varone, A., Verona, C.: Plasma carburizing of laser powder bed fusion manufactured 316 L steel for enhancing the surface hardness. Coatings. **12**, 258 (2022)

26. Bolli, E., Kaciulis, S., Lanzutti, A., Mezzi, A., Montanari, R., Palombi, A., Sordetti, F., Vaglio, E., Varone, A., Verona, C.: Surface characteristics and tribological behavior of 3D-printed 316 L steel after plasma assisted low temperature carburizing. Surface Coat. Technol. **477**, 130295 (2024)
27. Palombi, A., Montanari, R., Varone, A., Verona, C., Bolli, E., Kaciulis, S., Mezzi, A., Lanzutti, A., Sordetti, F., Vaglio, E.: 3D-printed 316 L stainless steel: optimization of low temperature plasma assisted carburizing. Metallurgia Italiana. **115**, 25–31 (2024)
28. Marinelli, M., Milani, E., Montuori, M., Paoletti, A., Paroli, P., Thomas, J.: High-quality diamond grow by chemical-vapor deposition: improved collection efficiency in -particle detection. Appl. Phys. Lett. **75**, 3216–3218 (1994)
29. Ridley, N.: Stuart H (1970) partial molar volumes from high-temperature lattice parameters of iron–carbon austenites. Metal Sci. J. **4**, 219–222 (1970)

Microstructural Stability of Co28Cr6Mo Biocompatible Alloy Produced by Additive Manufacturing

Alessandra Fava[1], Saulius Kaciulis[2], Alessio Mezzi[2], Roberto Montanari[3], Alessandra Palombi[3], and Alessandra Varone[3](✉)

[1] ENEA, Department for Sustainability—Research Centre of Casaccia, Santa Maria di Galeria, 00123 Rome, Italy
[2] ISMN-CNR, P.O. Box 10, 00015 Monterotondo Stazione, Rome, Italy
[3] Department of Industrial Engineering, University of Rome "Tor Vergata", Via del Politecnico 1, 00133 Rome, Italy
alessandra.varone@uniroma2.it

Abstract. Thanks to their excellent corrosion and wear resistance CoCrMo alloys are widely employed in the production of medical devices. Additive Manufacturing (AM) allows the fabrication of customized devices to fit the patient anatomy improving the implant functionality. AM gives rise to CoCrMo alloys with a fine cellular microstructure consisting of metastable γ phase and stable ε phase. This work investigated the microstructural stability of a Co28Cr6Mo alloy printed by Selective Laser Melting (SLM) after a heating-cooling cycle from room temperature to 800 °C. The results of X-ray diffraction (XRD) measurements, Light Microscopy (LM) observations and X-ray Photoelectron Spectroscopy (XPS) analyses show that part of the metastable γ phase transforms into the stable ε phase, and Mo and Cr diffuse towards the surface forming a stable oxide layer.

Keywords: Biocompatible alloys · CoCrMo · Additive manufacturing · Microstructural characterization

1 Introduction

The materials for the production of medical implants and surgery devices must guarantee specific mechanical properties, excellent corrosion resistance without harmful effects for the human body [1–4].

Among biocompatible materials, CoCrMo alloys are the most used in orthopaedic, orthodontic and dental applications thanks to their excellent mechanical properties, wear and corrosion resistance [5]. Their strong corrosion resistance is due to the spontaneous formation, in air, of a thin passive oxide layer. The passivation phenomenon consists in a decrease of the cobalt content on the surface and an increase of the thickness of the oxide layer. The beneficial effects of heat treatments (HT) on the corrosion resistance of CoCrMo alloys have been investigated by Zhou et al. [6]. The HT at 500 °C for 12 h reduces Co content on the surface and promotes the formation of the Cr and Mo oxides.

© The Author(s), under exclusive license to Springer Nature Switzerland AG 2025
M. Carfagni et al. (Eds.): IDBN 2024, BioSysRob 35, pp. 53–60, 2025.
https://doi.org/10.1007/978-3-031-94626-4_7

The presence of a thick oxide layer improves the corrosion resistance of the material and prevents the release of Co and Cr ions, dangerous for the human body.

The advent of Additive Manufacturing (AM) allows the production of highly tailored implants, that can better fit the patient anatomy, and overcome some of the manufacturing limits typical of Co-based alloys [7]. The result of AM processes, characterized by thermal cycles and high temperature gradients, is a microstructure with small grains, that presents a cellular substructure due to Mo segregation [8], and meta-stable phases [9]. Heat treatments (HT) are often performed on 3D printed parts to homogenize the microstructure and reduce residual stresses.

A lot of work has been done in the past to investigate the properties of CoCrMo alloys prepared through AM [9–12], both in as-built conditions and after post-process HT [13–15]. Recently, the microstructural evolution of the Co28Cr6Mo alloy prepared by SLM has been studied by us through Mechanical Spectroscopy (MS) tests carried out with heating ramps from room temperature up to 800 °C simulating HT [16]. MS measures dynamic modulus and damping giving information on the anelastic behaviour of a material [17–22]. The results evidenced the instability of interphase boundaries between γ and ε phases in as-built alloy and and surface Mo diffusion.

In this work the evolution of the microstructure and surface chemical composition has further investigated by Light Microscopy (LM) observations, X-ray diffraction (XRD) measurements and X-ray induced photoelectron spectroscopy (XPS) analyses.

2 Materials and Methods

The nominal chemical composition of the examined Co28Cr6Mo alloy is: Cr 27.0–30.0, Mo 5.0–7.0, Si < 1.0, Mn < 1.0, Fe < 0.75, Ni 0.25–0.50, C < 0.16, W < 0.20, Ti < 0.10, O < 0.10, Co to balance (wt.%). It was produced by SLM with a laser energy density of 119 J/mm^3.

The heat treatment has been performed inside the VRA instrument [23] in vacuum (10^{-6} mbar) from room temperature to 800 °C with a heating rate of 1.5 °C/minute.

XRD measurements were performed through a PW 1729 diffractometer (Philips, Eindhoven, The Netherlands) by using the Mo-Kα radiation, with a wavelength $\lambda =$ 0.07093 nm, 2θ steps of 0.05° and a counting time of 2 s per step. Phase identification was carried out by using the JCPDS database [24]. Quantitative analysis has been obtained, based on XRD patterns. In the case of a material with different phases the intensity of each peak of a single phase depends on the concentration C of the phase. Therefore, considering the intensities of every peak, the concentration of the γ and ε phases can be written as [25]:

$$C_\gamma = \frac{1}{a}\sum_{1}^{a}\frac{2\mu_m I_\gamma}{KR_\gamma} \quad (1)$$

$$C_\varepsilon = \frac{1}{b}\sum_{1}^{b}\frac{2\mu_m I_\varepsilon}{KR_\varepsilon} \quad (2)$$

where a and b are the number of γ and ε peaks in the examined pattern, K is a constant, μ the coefficient of linear absorption of the material and R is expressed by:

$$R = \left(\frac{1}{\nu^2}\right)\left[|F|^2 p\left(\frac{1+\cos^2 2\theta}{sen^2 2\theta\ cos\theta}\right)\right]\left(e^{-2M}\right) \quad (3)$$

(ν = volume of the unit cell, F = structure factor, p = multiplicity factor, θ = Bragg angle, e^{-2M} = temperature factor.)

As-built and heat treated samples were investigated by XPS. The analyses were performed by using an Escalab 250 Xi spectrometer (Thermo Fisher Scientific Ltd., East Grinstead, UK) equipped with a monochromatic Al X-ray source (hν = 1486.6 eV) and six-channeltron detection system. The analysed area of the samples was 900 × 900 μm^2 and the measurements were carried out operating at constant pass energy of 50 eV. The binding energy (BE) scale was calibrated positioning the adventitious carbon peak at BE = 285.0 eV. All the spectra were collected and processed by Avantage v.5 (Thermo Fisher Scientific Ltd., East Grinstead, UK).

3 Results and Discussion

Figure 1 shows the microstructure of the sample in as-built condition (a and b) and after HT (c and d). In the as-built sample it is possible to identify the melt pools (Fig. 1a) and a cellular sub-micrometric structure within them (Fig. 1b). After HT the melt pools disappear (Fig. 1c) and the material presents coarser microstructure with grains oriented along the building direction (Fig. 1d).

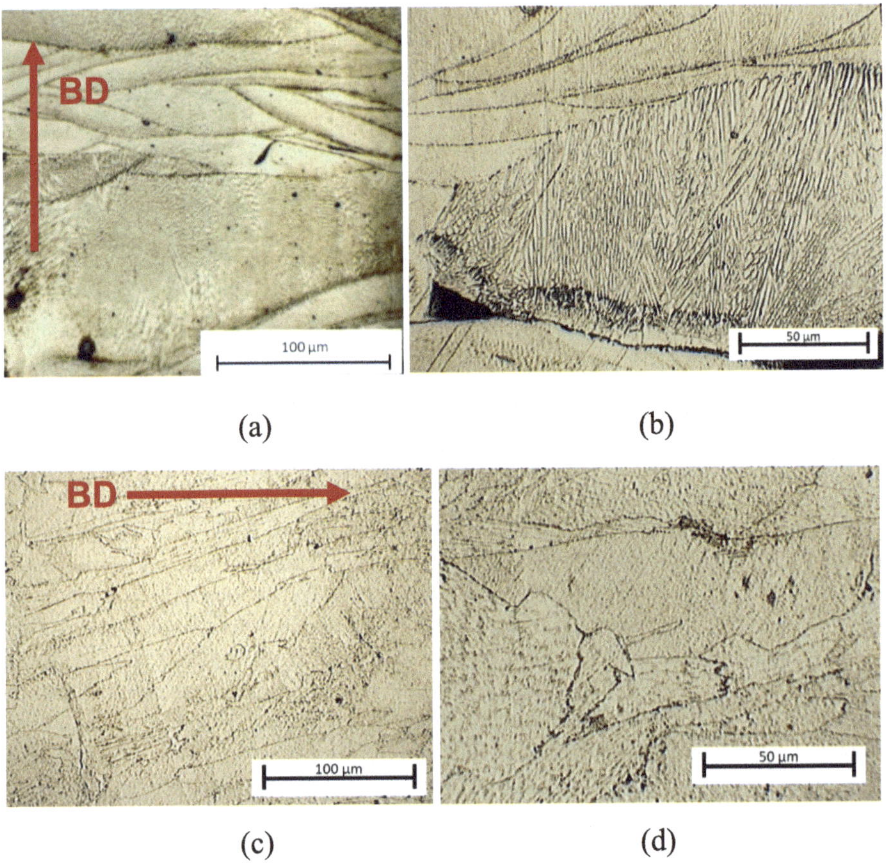

Fig. 1. Cross-section of as-built (a and b) and HT (c and d) sample at different magnification.

The as-built material is in metastable conditions, and the HT leads to an evolution in the microstructure which involves phase transformation (γ to ε), and grain re-orientation, as can be observed in the XDR patterns (Fig. 2). The ε phase is stable at room temperature, on the contrary γ is stable above 900 °C but is observed because the γ → ε transformation is slow and it has not enough time to occur due the fast cooling rates involved in SLM process. After the HT, new ε peaks are observed and their relative intensity increases compared to the as-built material.

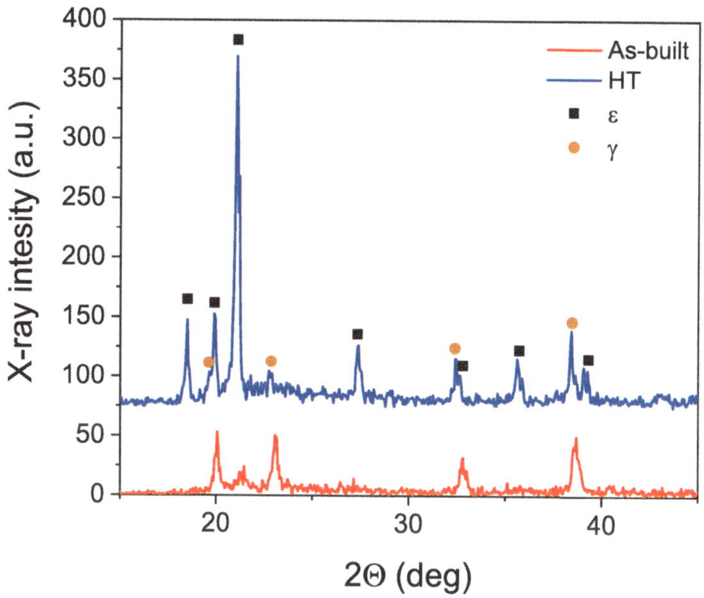

Fig. 2. XRD patterns of as-built and HT samples.

From the spectra reported in Fig. 2 a quantitative analysis has been carried out to determine the amount of γ and ε phases (Table 1).

Table 1. Amount of γ and ε phases in as-built and HT material.

	% γ	% ε
As-built	67	33
HT	45	55

To evaluate the effects of the HT on surface segregation of Mo and Cr, and, therefore, the formation of the surface oxide passive layer, XPS analysis has been employed.

As shown in Table 2, the relative amounts of Mo and Cr on the surface of the as-built sample are comparable with the nominal composition of the alloy. In AM processes, melted powders rapidly solidify due to high temperature gradients, therefore long-range diffusion is inhibited. This results in Mo and Cr concentrations of the as-built sample close to the nominal composition.

On the other hand, after HT the concentrations of Mo and Cr on the surface increase and result to be much higher than the nominal concentration. Of course, Co content exhibits the opposite trend. Therefore, HT favours the activation of diffusion processes of Mo and Cr towards the surface to reduce the Gibbs free energy. This elemental migration favors the microstructural evolution.

Table 2. Relative amounts (at.%) of Co, Cr and Mo at the surface for as built and HT material.

	Co	Cr	Mo
As-built	66.6	26.7	6.7
HT	40.7	50.0	9.3
Nominal composition	66.0	30.3	4.7

When in contact with oxygen, Mo and Cr rapidly form stable oxides: higher amounts of Cr and Mo on the surface lead to the formation of thicker oxide layers and thus the improvement in the corrosion resistance of the material.

Figure 3 shows the Cr2p and Mo3d core level peaks. The spectra reveal that the intensity of the peaks corresponding to the oxidized state, Cr_2O_3 and MoO_2, is higher in HT sample.

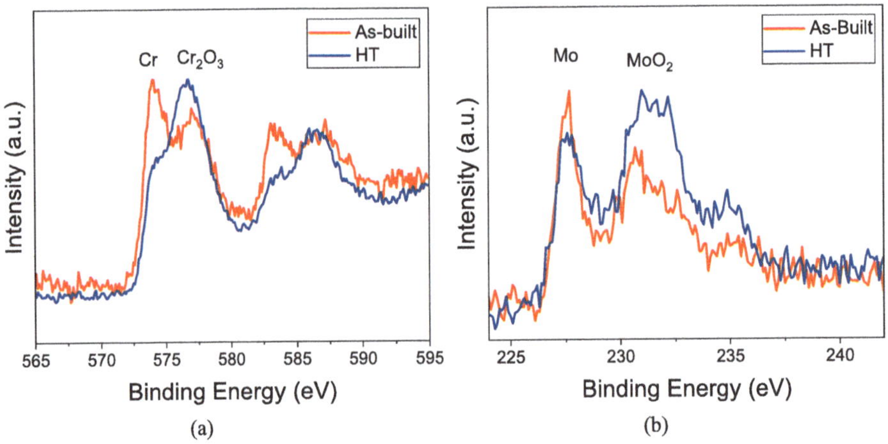

Fig. 3. Cr2p (a) and Mo3d (b) core level spectra on the surface of as-built and HT samples.

4 Conclusions

The results of present work can be resumed as follows.

(i) The Co28Cr6Mo alloy manufactured by SLM is not in conditions of thermodynamic equilibrium. The microstructure consists of small grains with a sub-micrometric cellular structure within them. Two phases are present, namely the stable ε and the metastable γ.

(ii) The heat treatment promotes of the transformation of γ phase into ε phase, that is accompanied by grain coarsening and texture change.

(iii) Following heat treatment the chemical composition of the surface drastically changes due to the migration of Cr and Mo towards the surface and Co in the opposite direction.

(iv) Cr and Mo form on the surface a thick layer of stable oxides, which should act as a barrier and improve corrosion resistance with beneficial effects for implants in contact with body fluids. Corrosion tests are underway to confirm this point.

Acknowledgements. The authors thank Prof. Lorella Ceschini of Bologna University for suppling the Co28Cr6Mo alloy used in present experiments.

References

1. Kiradzhiyska, D., Mantcheva, R.: Overview of biocompatible materials and their use in medicine. Folia Med. **61**, 34–40 (2019)
2. Poovathikkal, S., Nair, N.: Biomater. Biocompat., 161 (2018)
3. Manam, N.S., Harun, W.S.W., Shri, D.N.A., Ghani, S.A.C., Kurniawan, T., Ismail, M.H., Ibrahim, M.H.I.: Study of corrosion in biocompatible metals for implants: a review. J. Alloys Compd. **701**, 698–715 (2017)
4. Ambrogi, V., Bolli, E., Ceccarelli, M., Kaciulis, S., Mezzi, A., Montanari, R., Pakhomova, E., Richetta, M., Varone, A.: Surface modifications of biodegradable AZ31 alloy after immersion in physiological solution. Surf. Interface Anal. **55**, 474–479 (2023)
5. Mani, G.: Chapter 1b Metallic biomaterials: cobalt-chromium alloys. In: Murphy, W., Black, J., Hastings, G. (eds.) Handbook of Biomaterial Properties, pp. 159–166. Springer, New York (2016)
6. Zhou, K., Chen, J., Wang, T., Su, Y., Qiao, L., Yan, Y.: Effect of surface energy on protein adsorption behaviours of treated CoCrMo alloy surfaces. Appl. Surf. Sci. **520**, 146354 (2020)
7. Konieczny, B., Szczesio-Wlodarczyk, A., Sokolowski, J., Bociong, K.: Challenges of co–Cr alloy additive manufacturing methods in dentistry—the current state of knowledge (systematic review). Materials. **13**, 3524 (2020)
8. Acharya, S., Soni, R., Suwas, S., Chatterjee, K.: Additive manufacturing of co–Cr alloys for biomedical applications: a concise review. J. Mater. Res. **36**, 3746–3760 (2021)
9. Tonelli, L., Fortunato, A., Ceschini, L.: CoCr alloy processed by selective laser melting (SLM): effect of laser energy density on microstructure, surface morphology, and hardness. J. Manuf. Process. **52**, 106–119 (2020)
10. Takaichi, A., Suyalatu Nakamoto, T., Joko, N., Nomura, N., Tsutsumi, Y., Migita, S., Doi, H., Kurosu, S., Chiba, A., Wakabayashi, N., Igarashi, Y., Hanawa, T.: Microstructures and mechanical properties of Co–29Cr–6Mo alloy fabricated by selective laser melting process for dental applications. J. Mech. Behav. Biomed. Mater. **21**, 67–76 (2013)
11. Qian, B., Saeidi, K., Kvetková, L., Lofaj, F., Xiao, C., Shen, Z.: Defects-tolerant Co-Cr-Mo dental alloys prepared by selective laser melting. Dent. Mater. **31**, 1435–1444 (2015)
12. Tonelli, L., Boromei, I., Fortunato, A., Ceschini, L.: Selective laser melting of a CoCrMo alloy for biomedical applications: correlations between microstructure and process parameters. Metallurgia Italiana. **111**(6), 41–47 (2019)
13. Tonelli, L., Ahmed, M.M.Z., Ceschini, L.: A novel heat treatment of the additively manufactured Co28Cr6Mo biomedical alloy and its effects on hardness, microstructure and sliding wear behavior. Progr. Addit. Manuf. **8**, 313–329 (2023)
14. Cornacchia, G., Cecchel, S., Battini, D., Petrogalli, C., Avanzini, A.: Microstructural, mechanical, and Tribological characterization of selective laser melted CoCrMo alloy under different heat treatment conditions and hot isostatic pressing. Adv. Eng. Mater. **24**, 2100928 (2022)

15. Tonelli, L., Boromei, I., Liverani, E., Ceschini, L.: Microstructural evolution induced by heat treatment in the Co28Cr6Mo alloy produced by selective laser melting. Metallurgia Italiana. **113**(2), 22–30 (2021)
16. de Crescenzo C, Fava A, Montanari R, Palombi A, Varone A (2024) Mechanical spectroscopy of biocompatible metal alloys. In: Montanari R, Richetta M, Febbi M, Staderini EM (eds) Engineering Methodologies for Medicine and Sports. EMMS 2024. Mechanisms and Machine Science, 162. Springer, Cham. doi:https://doi.org/10.1007/978-3-031-63755-1_30
17. Nowick, A.S., Berry, B.S.: Anelastic Relaxation in Crystalline Solids. Academic (1972)
18. Golovin I (2007) Blanter-Golovin-Neuhauser-Sinning: Internal Friction in Metallic Materials.
19. Cabibbo, M., Montanari, R., Pola, A., Tocci, M., Varone, A.: Mechanical spectroscopy study of as-cast and additive manufactured AlSi10Mg. J. Alloys Compd. **914**, 165361 (2022)
20. Cabibbo, M., de Crescenzo, C., Fava, A., Montanari, R., Palombi, A., Pola, A., Tocci, M., Varone, A.: Dynamic modulus anomaly in metallic alloys prepared by additive manufacturing. Key Eng. Mater. **968**, 57–63 (2023)
21. Fava, A., Montanari, R., Varone, A.: Mechanical spectroscopy investigation of point defect-driven phenomena in a Cr martensitic steel. Metals. **8** (2018)
22. Bonetti, E., Cabibbo, M., Campari, E.G., Montanari, R.: Correlation between anelastic response and microstructure of 5N-Al thin foils. J. Alloys Compd. **872**, 159693 (2021)
23. Amadori, S., Campari, E.G., Fiorini, A.L., Montanari, R., Pasquini, L., Savini, L., Bonetti, E.: Automated resonant vibrating-reed analyzer apparatus for a non-destructive characterization of materials for industrial applications. Mater. Sci. Eng. A. **442**, 543–546 (2006)
24. JCPDS International Centre for Diffraction Data, Newtown Square, (n.d.)
25. Cullity, B.D.: Elements of X-Ray Diffraction, 2nd edn. Addison Wesley, Reading (1977)

Advanced Biomechanical Modelling Techniques and Materials' Analysis in Dental Structures

Giuliana Baiamonte[1(✉)], Fulvia Monaco[1], Giuseppe Laudani[1], Gianfranco Di Martino[1], Riccardo Asero[2], and Michele Calì[1]

[1] Electric, Electronics and Computer Engineering Department, University of Catania, 95125 Catania, Italy
{giuliana.baiamonte,fulvia.monaco,giuseppe.laudani, michele.cali}@unict.it, gianfranco.dimartino@phd.unict.it
[2] Asero Dental Clinic, Via Milano, 35, Misterbianco, CT, Italy
riccardoasero@me.com

Abstract. The increase in dental diseases and the high demand for dental aesthetics in developed countries have favoured the growth of demand in the dental market for restorative treatments for damaged teeth and beyond. Today, modern prosthetic dentistry aims to find new materials whose biomechanical behaviour approximates that of natural enamel when subjected to functional loads. Computer-aided design, engineering and manufacturing (CAD-CAE and CAD-CAM) can provide new technologies for the design and production of dental devices. The aim of this research is to analyse the bio-mechanical behaviour of new materials such as zirconia and silica-based ceramics; the effects of functional loads on dental prostheses are examined, starting with the acquisition of a lower first molar with a Cone Beam Computed Tomography (CBCT) scan and the subsequent CAD design of a 3D model for the restoration. These technologies will enable the fabrication of dental structures with solid materials and/or cell structures characterised by variable anisotropy and porosity, providing patients with stable results over time. This will help minimise waiting times and production costs, as well as provide customised solutions of high precision and reduced invasiveness.

Keywords: Dental devices · Basic reconstruction materials · Prosthetic crown · Mandibular kinematics · Biomechanical modelling

1 Introduction

The adoption of emerging digital technologies and the increasing rise of dental diseases are profoundly modifying the techniques of dental care, replacing manual processes with updated procedures; the latter are driving the growth of demand in the dental market in developed countries to carry out repairing treatments for damaged teeth and it is estimated that between 2021 and 2027 their annual growth rate will be 5.2% [1]. A tooth fracture is a significant problem for both the patient and the dentist and often a large percentage of fractured teeth cannot be restored; this involves breaking down the structure

© The Author(s), under exclusive license to Springer Nature Switzerland AG 2025
M. Carfagni et al. (Eds.): IDBN 2024, BioSysRob 35, pp. 61–68, 2025.
https://doi.org/10.1007/978-3-031-94626-4_8

of the tooth, exposing sensitive internal tissue, such as pulp and nerves, causing discomfort and pain to the subject. Fractures can be caused by a variety of factors, including untreated cavities, weakening of the tooth following invasive dental procedures, or significant force exerted on hard objects. The masticatory muscles, i.e. masseter, temporal and external-internal pterygoid, in fact, exert forces that vary in a range from 600 to 1200 N on all teeth and on both sides depending on the food interposed between them. The posterior teeth are the most prone and in particular the first mandibular molars, in which fractures are more frequent in the lingual cusps than in the buccal ones. To date, the goal of modern prosthetic dentistry is to find new materials to carry out restorative dental treatments, whose biomechanical behavior is close to that of natural enamel when subjected to functional loads [2]. Proper selection of restorative materials is critical to minimize restorative wear and improve tissue compatibility while ensuring the effectiveness of the masticatory system [3, 4]. With the development of ceramic materials for oral rehabilitation, satisfactory aesthetic results have been achieved, particularly in the anterior region, because they allow to achieve a high resemblance to the natural tooth [5, 6]; however, ceramic materials are brittle with low mechanical strength, making it susceptible to cracks and fractures during its use [7, 8]. Recently, zirconia (ZrO_2) has been considered a base material for restorations, as it has high mechanical and corrosive strength, lower magnetic susceptibility and lower cytotoxicity [9]. In this context, computer-aided design and manufacturing (CAD-CAM), together with 3D printing techniques, can provide new technologies for dental devices, reducing tooth treatment and fabrication procedures [10, 11]. In particular, multiple 3D printing techniques allow dentists to use different materials: Selective laser sintering (SLS) for printing powders, stereolithography (SLA) for photopolymers (ceramics and filled resins), fused deposition modelling (FDM) for thermoplastics and digital light processing (DLP) for printing various materials to repair damaged dental structures [12, 13]; polymers are considered to be well-suited for additive manufacturing methods in creating crowns, denture bases and artificial teeth [14]. The fully digital workflow starts from a few minutes scan in order to obtain a 3D model and send it to a 3D printer to print a detailed dental prosthesis in a same-day fabrication. Moreover, the 3D printers influence the dental restorations' accuracy and this problem was addressed by Vichi A. et al. in order to identify the correct angle of positioning in different restoration types [15]. Nowadays, the use of resins low in methacrylate and light-curing resins [16–18] is also envisaged; Molded resins fit different specifications and clinical needs and are less expensive than milled ceramics. In this case, the most commonly used manufacturing method is tank polymerization, in which liquid resins are partly cured in one tank and then totally cured in another polymerization unit [19, 20]. The present research aims to analyze the biomechanical behavior of materials, such as zirconia and silica-based ceramics, with reference to previous works conducted by the authors on topics related to the analysis, reconstruction, printing of dental implants [21] and modeling of the mandibular bone [22]; the effects of functional loads on dental support materials will be examined through the development of the phases related to the acquisition of a lower first molar and the design of a 3D model of the tooth and a crown for restoration with a CAD software, in order to compare their mechanical characteristics. These technologies will make it possible to design dental structures with solid materials and/or materials with cellular structures characterized by variable anisotropy and porosity [23], providing the patient with stable results over time. This will help minimize waiting times and production costs, as well as provide customized solutions with high accuracy and reduced invasiveness. The rest of the manuscript is divided as follows: Section 2 describes the proposed methodology; in Section 3 the discussions and the main results obtained were reported and finally in Section 4 the conclusions of the research were drawn.

2 Materials and Methods

In the development of the present research work, three phases can be distinguished: initially, the acquisition of the upper and lower arches of the teeth with normal occlusion is planned by scanning with CBCT and segmentation is performed, obtaining a first lower molar; in the second phase, the 3D CAD reconstruction of the geometries involved is carried out and finally the finite element analysis (FEA) is conducted. The methods and tools used in these three steps are described in Fig. 1 and in the following subsections.

2.1 CBCT Data Analysis and Segmentation

The first phase of the proposed methodology involved the acquisition of the upper and lower arches of the teeth of patients with normal occlusion by CBCT scan (Fig. 2 a, b). These files in DICOM format have been imported into open-source software, in order to extract information about the teeth and in particular about the size of the lower first

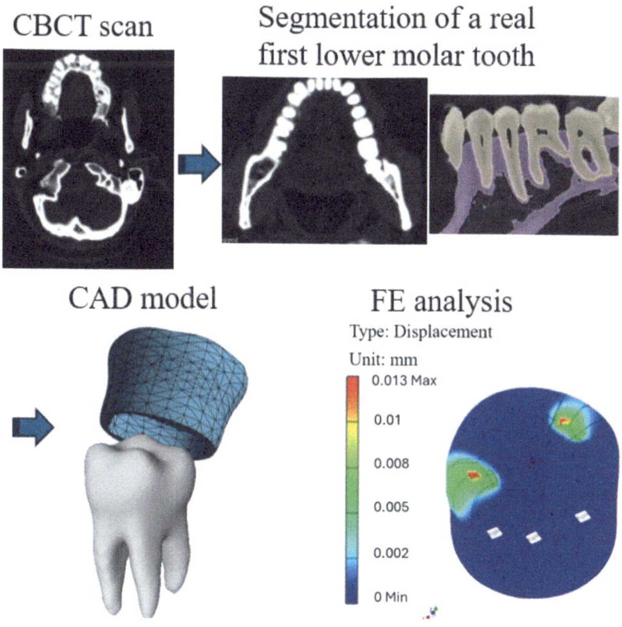

Fig. 1. Flowchart of the developed methodology

molar; using the segment editor tool, the mandible was analyzed by creating a region of interest (ROI) to select the volume to be segmented accordingly. This made it possible to create a mask and to perform segmentation on some slices, obtaining the dimensions of the tooth under study (Fig. 2c).

2.2 Prosthetic Crown 3D CAD-CAE Model

The CAD model of the the prosthetic crown was obtained using the Autodesk Inventor 2024 CAD environment software with a thickness ranging from 0.3 mm to 1.5 mm. Subsequently, a cherry pit with dimensions of 8 mm in length and 4 mm in diameter was also modelled to simulate the possible contact points in specific areas between it and the prosthetic crown during chewing, in order to evaluate the bio-mechanical behaviour of the two materials. The 3D models of these elements were discretized with tetrahedral elements (Figs. 3a, b), where the number of nodes and elements applied to each part is shown in Table 1. In order to analyse the mechanical behaviour of the model, the two basic materials used for dental restoration, ZrO_2 and silica-based ceramics, were considered and in Table 2 the mechanical properties of the materials are shown.

It was assumed that all the materials have an isotropic behaviour with the same thermo-mechanical characteristics in all directions and only the stresses and deformations of the upper part of the tooth were analyzed. Some authors in their research works, reported by Mehri Sofiani, F. et al. [24], determined a force range of = 25 to 450 N and

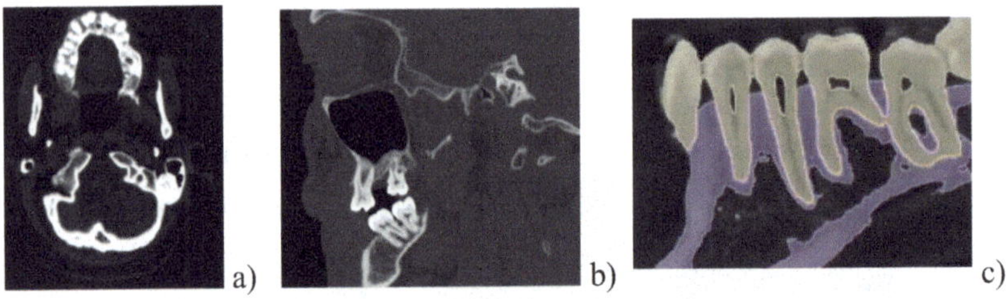

Fig. 2. CBCT scan of the mandible: a) view of the axial plane; b) view of the sagittal plane; c) segmentation process of the lower first molar

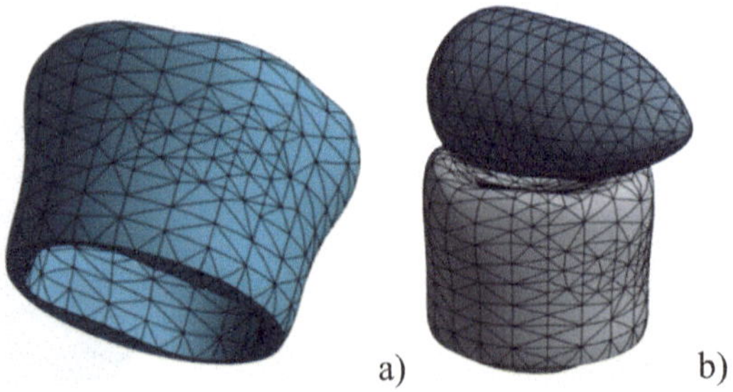

Fig. 3. Discretized parametric models: a) prosthetic crown; b) prosthetic crown with cherry pit on top

Table 1. FE Analysis: nodes and tetrahedral elements.

Component	n. of elements	n. of nodes
Prosthetic crown	2240	4587
Cherry pit	1645	2985

Table 2. Mechanical properties of the materials included in finite element analysis.

Component	Material	Elastic modulus (GPa)	Poisson ratio	Tensile Strength (MPa)	Compressive Strength (MPa)	Yield stresses (MPa)
Tooth	Enamel	84.1	0.33	11.5	384	–
	Dentin	18	0.31	105.5	297	–
Prosthetic crown	Zirconia	205	0.3	745	2000	230
	Silica-based ceramics	69 ÷ 70	0.3	20 ÷ 60	350 ÷ 550	560
Cherry pit	–	0.19	0.25	–	–	–

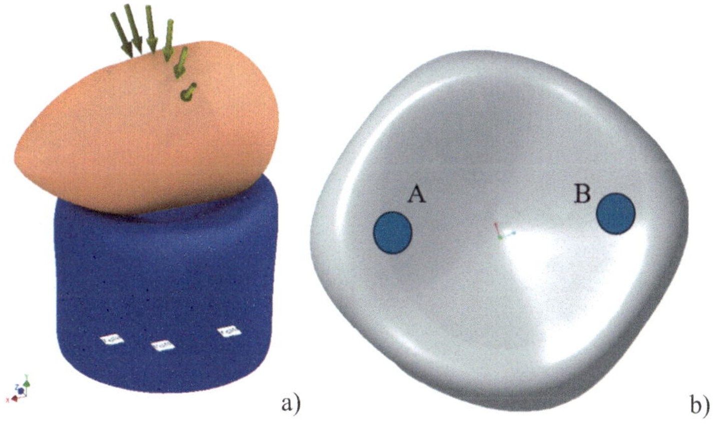

Fig. 4. a) Distributed load on the cherry pit in contact with the prosthetic crown; b) load's application zones

= 100 to 500 N as an occlusal load to investigate the effect of occlusal restoration and loading on the development of abfraction lesions through finite element method (FEM). In this case study a distributed load with a resultant of 700 N, considering vertical and inclined components, was applied to the cherry pit to simulate contact during chewing. In Fig. 4 a and b the distribution of forces and the contact zones A and B on the occlusal surface of the prosthetic crown are visible.

3 Results and Discussion

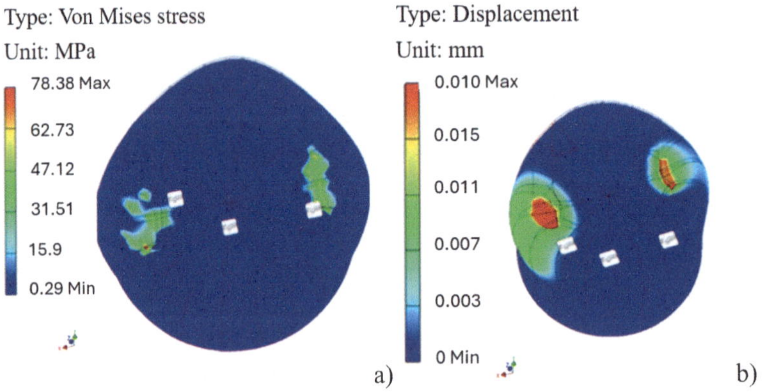

Fig. 5. a) Von Mises stress distribution; b) displacements in occlusal zone

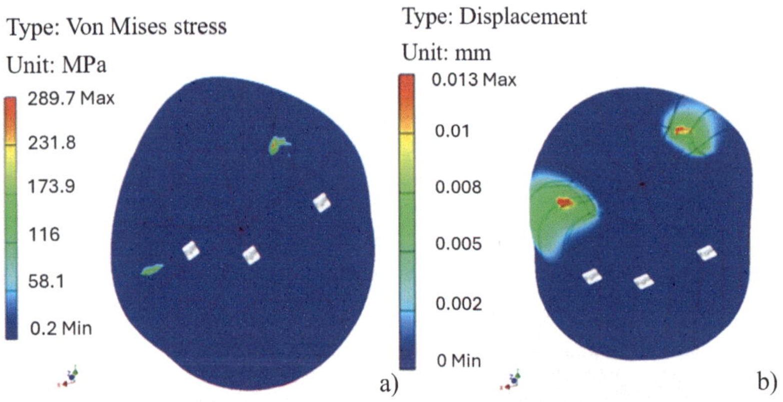

Fig. 6. a) Von Mises stress distribution; b) displacements in occlusal zone

Figures 5 a,b and 6 a,b show the results of the FEM analysis conducted on both types of material applied to the prosthetic crown. The study found that in the occlusal area of the prosthetic crown, and in particular in points A and B, greater stress is concentrated; in the first case, in which the prosthetic crown is made with ZrO_2, a maximum Von Mises stress of 78.38 MPa and a maximum displacement of about 0.01 mm were found; in the second case, using silica-based ceramics, a maximum stress value of 289.7 MPa and a maximum displacement of 0.013 mm were obtained. This made it possible to analyze the compressive strength of both materials on the two ends of the prosthetic crown where greater contact occurs, highlighting how both dental reconstruction materials are highly resistant and differ only in the biocompatibility factor, as well as aesthetics and construction methods. In addition, the highest stresses and deformations occurred only in the occlusal area of the prosthetic crown without propagating to the vestibulo-lingual area.

4 Conclusions

This research work focused on the bio-mechanical strength of two types of base materials, zirconia and silica-based ceramics, used for dental restorations such as for crowns, bridges and implants. Using finite element analysis on a 3D CAD model of the prosthetic crown and including a food bolus with linear and elastic characteristics, i.e. the cherry pit, a distributed load of 700 N was applied to simulate a real chewing situation, as the molars are more subject to high forces than the other teeth. Future studies plan to analyze other case studies, interposing a bolus of true size and mechanical characteristics between specific teeth, in order to get closer and closer to the real masticatory function. Additional forces acting on the teeth will be evaluated, integrating the components that occur during the circular movement of chewing, varying them depending on the texture of the food. It is plan to conduct studies on how to modify the geometries of restorations (e.g., crowns, bridges) to improve stress distribution and reduce critical points of high stress; the evaluation of materials' behaviour under repeated stress over time will be analyse, considering fatigue and wear phenomena and conditions of prolonged use. Moreover, it is plan to integrate FEA results with advanced imaging techniques, such as computed tomography (CT) and magnetic resonance imaging (MRI), to gain a more detailed view of material behaviour.

References

1. https://www.gminsights.com/industry-analysis/dental-cad-cam-market
2. Wilmers, J., Bargmann, S.: Nature's design solutions in dental enamel: uniting high strength and extreme damage resistance. Acta Biomater. **107**, 1–24 (2020)
3. Oblak, C., Kocjan, A., Jevnikar, P., Kosmac, T.: The effect of mechanical fatigue and accelerated ageing on fracture resistance of glazed monolithic zirconia dental bridges. J. Eur. Ceram. Soc. **37**(14), 4415–4422 (2017)
4. Borrero-Lopez, O., Guiberteau, F., Zhang, Y., Lawn, B.R.: Wear of ceramic-based dental materials. J. Mech. Behav. Biomed. Mater. **92**, 144–151 (2019)
5. Frasnelli, G.D., Leoni, G.B., Cassani, R., Faria, A.C.L., Gomes, E.A., Moris, I.C.: Effect of different ceramic systems on antagonist dental structure by microtomographic analysis. Dent. Mater. **40**, 118–123 (2023)
6. Shi, H.Y., Pang, R., Yang, J., Fan, D., Cai, H., Jiang, H.B., et al.: Overview of several typical ceramic materials for restorative dentistry. Biomed. Res. Int. **2022**(1), 8451445 (2022)
7. Pieger, S., Salman, A., Bidra, A.S.: Clinical outcomes of lithium disilicate single crowns and partial fixed dental prostheses: a systematic review. J. Prosthet. Dent. **112**(1), 22–30 (2014)
8. Øilo, M., Hardang, A.D., Ulsund, A.H., Gjerdet, N.R.: Fractographic features of glass-ceramic and zirconia-based dental restorations fractured during clinical function. Eur. J. Oral Sci. **122**(3), 238–244 (2014)
9. Chopra, D., Guo, T., Gulati, K., Ivanovski, S.: Load, unload and repeat: understanding the mechanical characteristics of zirconia in dentistry. Dent. Mater. **40**, e1–e17 (2023)

10. Della Bona, A., Pecho, O.E., Alessandretti, R.: Zirconia as a dental biomaterial. Materials. **8**(8), 4978–4991 (2015)
11. Bagheri, H., Hooshmand, T., Aghajani, F.: Effect of ceramic surface treatments after machine grinding on the biaxial flexural strength of different CAD/CAM dental ceramics. J. Dentist. (Tehran, Iran). **12**(9), 621–629 (2015)
12. Rezaie, F., Farshbaf, M., Dahri, M., Masjedi, M., Maleki, R., Amini, F., et al.: 3D printing of dental prostheses: current and emerging applications. J. Composit. Sci. **7**(2), 80 (2023)
13. Sahai, N., Sunil, B.D.Y., Chandrashekar, R.: 3D printing, modelling and simulation for dental implant: a comprehensive review. Innov. Emerg. Technol. **11**, 2430001 (2024)
14. Dimitrova, M., Vlahova, A., Kalachev, Y., Zlatev, S., Kazakova, R., Capodiferro, S.: Recent advances in 3D printing of polymers for application in prosthodontics. Polymers. **15**(23), 4525 (2023)
15. Vichi, A., Balestra, D., Scotti, N., Louca, C., Paolone, G.: Translucency of CAD/CAM and 3D printable composite materials for permanent dental restorations. Polymers. **15**(6), 1443 (2023)
16. Saini, R.S., Gurumurthy, V., Quadri, S.A., Bavabeedu, S.S., Abdelaziz, K.M., Okshah, A., et al.: The flexural strength of 3D-printed provisional restorations fabricated with different resins: a systematic review and meta-analysis. BMC Oral Health. **24**(1), 66 (2024)
17. Della Bona, A., Cantelli, V., Britto, V.T., Collares, K.F., Stansbury, J.W.: 3D printing restorative materials using a stereolithographic technique: a systematic review. Dent. Mater. **37**(2), 336–350 (2021)
18. Alshadidi, A.A.F., Alshahrani, A.A., Aldosari, L.I.N., Chaturvedi, S., Saini, R.S., Hassan, S.A.B., et al.: Investigation on the application of artificial intelligence in prosthodontics. Appl. Sci. **13**(8), 5004 (2023)
19. Bora, P.V., Sayed Ahmed, A., Alford, A., Pitttman, K., Thomas, V., Lawson, N.C.: Characterization of materials used for 3D printing dental crowns and hybrid prostheses. J. Esthet. Restor. Dent. **36**(1), 220–230 (2024)
20. Andjela, L., Abdurahmanovich, V.M., Vladimirovna, S.N., Mikhailovna, G.I., Yurievich, D.D., Alekseevna, M.Y.: A review on vat photopolymerization 3D-printing processes for dental application. Dent. Mater. **38**(11), e284–e296 (2022)
21. Calì, M., Zanetti, E.M., Oliveri, S.M., Asero, R., Ciaramella, S., Martorelli, M., Bignardi, C.: Influence of thread shape and inclination on the biomechanical behaviour of plateau implant systems. Dent. Mater. **34**(3), 460–469 (2018)
22. Pascoletti, G., Aldieri, A., Terzini, M., Bhattacharya, P., Calì, M., Zanetti, E.M.: Stochastic PCA-based bone models from inverse transform sampling: proof of concept for mandibles and proximal femurs. Appl. Sci. **11**(11), 5204 (2021)
23. Latimer, J.M., Maekawa, S., Yao, Y., Wu, D.T., Chen, M., Giannobile, W.V.: Regenerative medicine technologies to treat dental, oral, and craniofacial defects. Front. Bioeng. Biotechnol. **9**, 704048 (2021)
24. Mehri Sofiani, F., Farahani, B.V., Belinha, J.: Elasto-static analysis of composite restorations in a molar tooth: a meshless approach. Polymers. **16**(4), 458 (2024)

Development and Testing of a Patient-Specific Cardiovascular Simulator for Interventional Valve-in-Valve Procedure Training

Emanuele Gasparotti[1(✉)], Benigno Marco Fanni[1], Roberta Zola[2], Chiara Corno[2], Massimiliano Mariani[3], Sergio Berti[3], and Simona Celi[1]

[1] BioCardioLab, Bioengeneering Unit, Fondazione Toscana "G. Monasterio", Massa, Italy
{gasparotti,bmfanni,s.celi}@ftgm.it

[2] Edwards Lifescience, Irvine, USA

[3] Adult Cardiology, Fondazione Toscana "G. Monasterio", Massa, Italy
https://bcl.ftgm.it/

Abstract. Patients with degeneration of a prosthetic mitral valve can be treated with minimally invasive transcatheter mitral valve in valve (TMViV) intervention, guided via fluoroscopic and echocardiographic intra-procedural imaging. However, despite the significant improvements in treatment outcomes and patient safety brought by the TMViV intervention, this procedure involves several steps requiring a highly skilled and experienced operator to be performed safely. Simulation-based procedural training has become a valuable method for efficiently facilitating the learning process for young operators across various types of interventions. The aim of this study was to present a new high-fidelity simulator featuring patient-specific anatomy and to demonstrate its potential impact on the procedural training for TMViV. The simulator was based on computed tomography to replicate the anatomy of a real patient using additive manufacturing. The intra-procedural imaging view were implemented with an ad-hoc positioning and orientation of specific cameras to reproduce the structures and anatomical landmarks displayed during the procedure via fluoroscopy and echocardiography. The entire TMViV procedure was simulated with real clinical instrumentation, including prosthetic and implanted valves, by expert operators demonstrating the feasibility of the simulator to be used as a realistic replicator of the anatomies involved during the intervention and the intra-procedural imaging views. The simulator represents a significant advancement in training capabilities, potentially improving procedural outcomes and patient safety in TMViV interventions.

Keywords: Mitral regurgitation · transcatheter mitral valve-in-valve procedure · 3D printing · Cardiovascular simulator training · Transseptal puncture · TAVI

© The Author(s), under exclusive license to Springer Nature Switzerland AG 2025
M. Carfagni et al. (Eds.): IDBN 2024, BioSysRob 35, pp. 69–80, 2025.
https://doi.org/10.1007/978-3-031-94626-4_9

1 Introduction

The standard treatment for patients with severe mitral valve disease is surgical valve replacement [7,13]. However, implanted valve degeneration may lead to many complications, such as need of a new surgical valve replacement which is associated with major complications including death [5]. For this reason, clinical interest in minimally invasive alternatives continues to grow. Transcatheter mitral valve-in-valve (TMViV) procedure has emerged as an optimal treatment for high-risk patients who require replacement of the degenerated valve [11,12]. TMViV intervention is performed under the transesophageal echocardiography (TEE) and fluoroscopic guidance, by exploiting a transcatheter aortic valve implantation (TAVI) in the mitral position.

During the procedure, the new bioprosthetic valve is passed via femoral venous access and introduced through transseptal puncture (TSP) of the fossa ovalis (FO) in the left atrium (LA), where it is finally expanded onto the ring of the degenerated prosthesis.

Despite the significant improvements in treatment outcomes and patient safety brought about by the TMViV intervention, the procedure complexity, such as navigating and positioning within the heart and the precise deployment of the new valve, necessitates a highly skilled and experienced operator [13]. In particular, the capability of handling the clinical instrumentation under the intra-operative image guidance plays a crucial role for a successful procedure. Hence, the training of TMViV procedure is fundamental to speed up the learning curve of new operators, especially considering the importance of TSP and its influence on the positioning of the new valve [7,15]. Simulation-based procedural training started to represent a valuable way to efficiently ease the learning process of young operators for different types of interventions [9], thus enhancing skill acquisition and ensuring thorough preparation for complex medical procedures [2]. The purpose of this study was to describe a novel high-fidelity simulator with patient-specific anatomy capable of simulating the entire TMViV procedure with real clinical instrumentation and to demonstrate its potential impact on procedural training.

2 Materials and Methods

The simulator was designed and developed to fulfill (*i*) the geometrical requirements to represent a real patient's anatomy and also (*ii*) the condition of having a realistic replica of the main intra-procedural guidance imaging views during the TMViV simulations. Requirement (*i*) was accomplished by using the computed tomography (CT) images of a patient to obtain the heart model of the simulator. For requirement (*ii*), realistic views of the intra-procedural imaging were implemented by opportunely positioning and orienting specific cameras to replicate the same visual information available in the cath lab during the intervention. Both requirements were fundamental to allow the exact execution of all the interventional steps of the TMViV [7,14], which are:

- insert the transseptal introducer (TI) with its dedicated guidewire into the inferior vena cava up to the superior vena cava;
- remove the guidewire and introduce the Brockenbrough (BRK) needle into the transseptal introducer;
- position the transseptal introducer at the inferior-posterior region of FO under the guidance of TEE bicaval and short axis views;
- proceed with septum puncture, advancing the BRK needle into FO;
- advance a stiff wire into the left superior pulmonary vein;
- remove the transseptal system and position the steerable introducer catheter;
- advance a pigtail catheter through the introducer catheter and cross the mitral valve prosthesis;
- through the pigtail, advance and deploy a spiral pre-shaped wire in the left ventricle;
- insert the delivery system sheath into the right femoral vein;
- perform balloon septostomy;
- insert the TAVI delivery system with collapsed valve into the delivery system;
- proceed with valve alignment in the delivery system;
- insert the delivery system across the septum;
- position the TAVI across the degenerated prosthesis;
- start balloon inflation for the TAVI deploy;
- deflate the balloon and remove the delivery system;
- remove the pre-shaped wire and the pigtail catheter.

The simulator was developed to capture all the interventional phases summarized above. The details of the developed simulator and its testing are given in the following sections, including image processing to obtain the geometries, design, fabrication of the system components, and TMViV simulations.

2.1 Image Processing

The anatomy of the simulator was modeled starting from a patient-specific CT dataset, selected based on clinical expertise to be representative of an average subject. A semi-automatic segmentation approach was implemented to identify the structures of interest related to the left heart (LH) and right heart (RH). The software Slicer 3D [8] was used for the segmentation. The final anatomical model for the simulator consisted in: the iliac veins (IVs) up to the lumbar veins; the right atrium (RA) with inferior (IVC) and the superior (SVC) venae cavae; the region of the FO; the LA including pulmonary veins (PVs) and the annulus of the MV; the left ventricle (LV) and the aortic root.

2.2 Design

The simulator was designed to replicate the anatomical structures involved during the TMViV and the intra-procedural imaging views as they occur during the real intervention. The system was composed of five anatomical parts: the femoral veins, the RA, the FO, the LA and the LV including aortic root. These components were assembled to replicate the same relative position and orientation as for a real patient lying on the cath lab table. The anatomical parts were referred to a rigid reference plane, representing the operating table, by means of specific supports (S).

The geometries of the segmented anatomies were thickened of 1.5 mm using SpaceClaim software (Ansys, United States), opened in specific regions to allow the acquisition of procedural imaging, and modified to allow their mutual coupling and a rapid assembly/disassembly with the supports. In particular, the exact alignment between the IVs and the IVC was guaranteed by their corresponding supports, i.e., S_1, S_2 and S_3 for the IVs, and S_4 and S_5 for the RA (Fig. 1a-b). The LH component, constituted by the LA and LV, was designed for an easy and rapid assembly assured by a slider/rail system (SL/R_1) which is hold in position through an apposite stopper (St) (Fig. 1b-c). LA and LV were connected by two bolts and a specific seat was designed at the MV annulus to firmly house a prosthetic surgical mitral valve that can be easily disassembled and replaced as needed.

Regarding the implementation of the intra-procedural imaging, four specific cameras (C) were adopted, positioned to accurately reproduce the same views visualized via XA and TEE. In particular, the views reproduced by the cameras were: the short-axis (C_1) and bicaval (C_2) views given by standard TEE, which are needed to orientate the needle in the right posterior-inferior position of the FO for the TSP and to guide the septostomy balloon; the surgical view (C_3) for the visualization of the MV from the LA, obtained from 3D TEE, for a correct centering of the new valve; the fluoroscopic coplanar view (C_4) to guide the catheter positioning in the LV and to set the centering and the height positioning of the deployed valve. Each camera was oriented to implement the corresponding simulated intra-procedural view by designing specific supports (SC_1, SC_2, SC_3 and R_2, depicted in Fig. 1a-b.

The FO model was designed as a separable component to be attached on the corresponding zone in the RA model. Six circumferential holes were created on the FO part, which is attached by inserting them into the corresponding pins (Pn) in the outer wall of the RA. Three independent holders (H) were realized to firmly hold the FO in position (Fig. 1e-d), allowing its rapid disassembly, thus allowing a safe passage of the clinical instrumentation during all the procedural phases without any risk of dislodgement.

A complete representation of the simulator design is reported in Fig. 1, including lateral (Fig. 1a) and top (Fig. 1b) views, and details of FO hooking system (Figs. 1c-d) and LA/LV system for the housing of the valve (Fig. 1e).

2.3 Fabrication Processes and Component Selection

The components of the simulator were fabricated with different techniques and materials. The reference plane base was made of PMMA. The anatomical parts were manufactured using 3D printing and casting techniques. In particular, the RA, including also the holders H, the LA and the LV were 3D printed with the transparent Clear resin (Formlabs, United States) by means of stereolitography

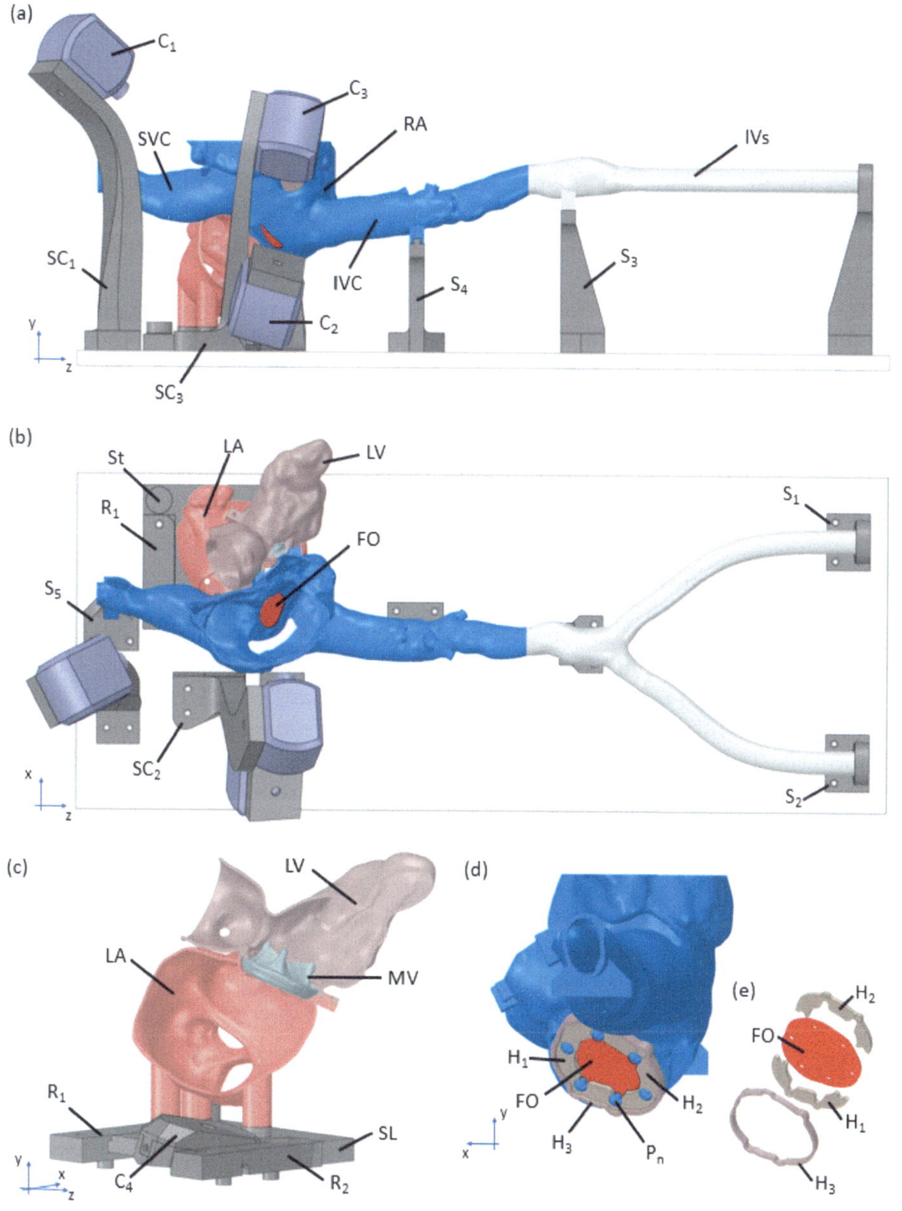

Fig. 1. Designed simulator views: lateral view (a), top view (b), left heart section with prosthetic mitral valve (c) and frontal view of RA with detail of FO and holder components (d) and the explosion of FO holders assembly (e).

(SLA) technology. These components were then coated with a bi-component glossy paint to increase their transparency and guarantee compliance with the camera views. The IVs model was created using fused deposition modeling (FDM), by printing white ABS with an A4V4 machine (3ntr, Italy). All the support components, including S, SC, R and St, were manufactured in SLA with Grey resin (Formlabs, United States).

The FO was instead fabricated through a casting procedure, by pouring the Ecoflex 00–30. A nylon net (denier 10, fibre diameter 43-45 μm, porosity 50%) was embedded during the pouring process to increase the resistance of the FO to the multiple passages of the clinical instrumentation through the hole of the TSP, during the procedure.

Regarding the hardware for implementing the intra-procedural image views, three Marshall CV506 cameras (Marshall Electronics, United States) and a Logitech B525 HD webcam (Logitech, Switzerland) were adopted. In particular, the Marshall cameras (C1, C2 and C3) were adopted to reproduce the standard TEE views (short-axis and bicaval) and the fluoroscopic coplanar view, while the webcam (C4) captured the 3D TEE surgical view. The cameras were connected to as many portable monitor for the display of the intra-procedural views.

2.4 TMViV Simulations

Two experts in TMViV were in charge of executing the simulations of the procedure. A Perimount plus surgical mitral valve (Edwards Lifesciences, United States) was adopted to reproduce the degenerated prosthesis. A Fast-Cath introducer sheath (Abbott, United States) was used to introduce the BRK needle and perform the TSP. Following the TSP, a Agilis NxT steerable introducer (Abbott, United States) was used to reach the LV and get across the prosthetic mitral valve. A Safari^2 pre-shaped guidewire (Boston Scientific, United States) was inserted through a pigtail catheter in the LV. At this point, the septostomy was performed using a PTA balloon dilatation catheter (Cook Medical, United States). The Edwards Commander system was adopted to deliver in the mitral position a SAPIEN 3 valve (Edwards Lifesciences, United States), which was inflated with the embedded balloon.

3 Results

The fabrication of the simulator was feasible as designed. 3D printing of the modeled anatomical parts was successful despite the high geometrical complexity including cavity and numerous hollowed parts (Fig. 2a). The simulator was designed to be translucent, enabling clear visibility through the walls to observe the clinical instrumentation (Fig. 2b). Additionally, strategically placed windows provided operators with a comprehensive view of all anatomical structures involved in the procedure. Moreover, these openings allowed for a realistic implementation of the intra-procedural views. The assembly of the anatomical components on the reference plane resulted fast, easy and repeatable, as well

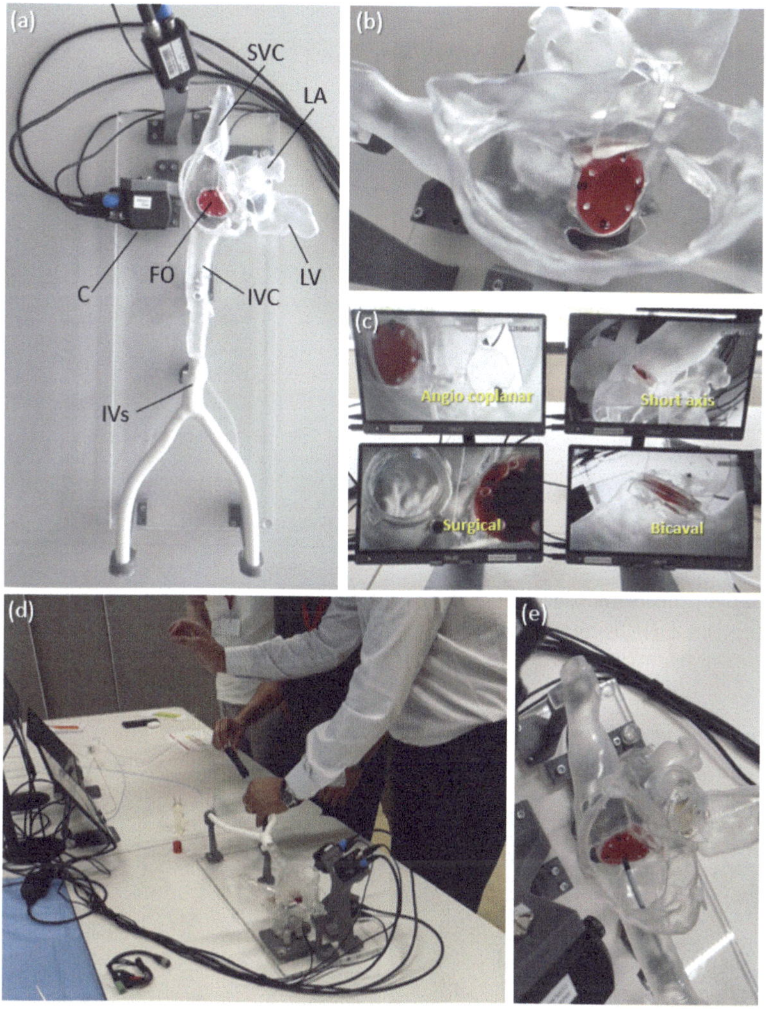

Fig. 2. Manufactured simulator: top view of the anatomical components and the cameras (a), zoom of the RA with the FO (b), camera views projected on the dedicated monitors (d), pictures of the simulator while used by clinicians (d-e).

as the replacement of the FO and the surgical valve between simulations. The feedback of the catheters introduced in the simulator results similar to the ViV procedure. The selection of a silicone and nylon net compound for creating the FO tissue was highly effective, providing operators with reliable tactile feedback during TSP execution and instruments introduction. Specifically, the modeled FO allowed for puncturing, septostomy, and introduction of catheters and valve delivery system without incurring damage.

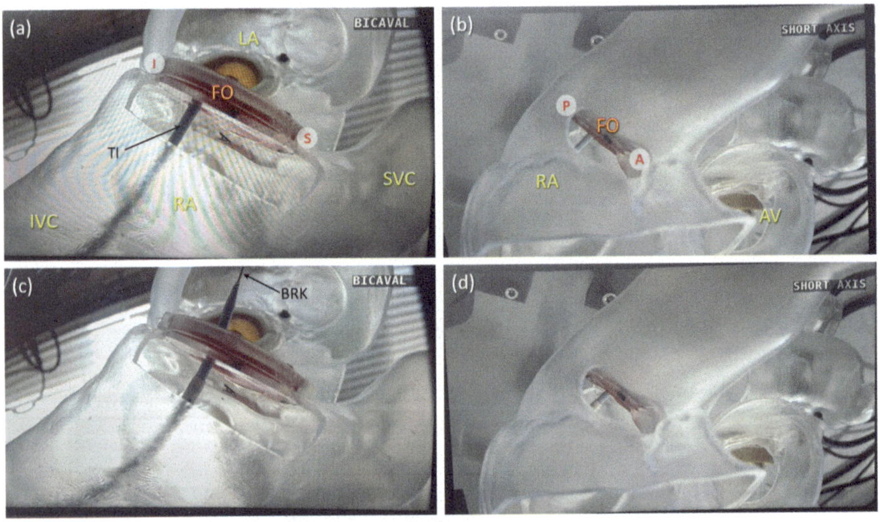

Fig. 3. TSP procedure: assessment of the inferior-posterior (I-P) position of TI on the FO before the TSP from bicaval (a) and short axis (b) TEE views; TI and BRK after FO puncture from the bicaval (c) and short axis (d) views.

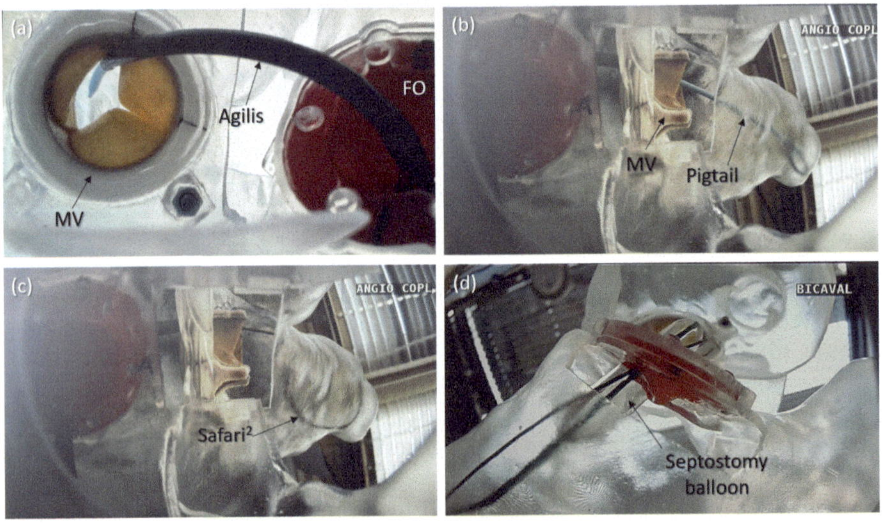

Fig. 4. Example of TMViV intermediate steps: crossing of Agilis and pigtail catheter through the mitral valve prosthesis from surgical view (a); angiographic coplanar view during the positioning of pigtail in the LV (b) and deployment of Safari^2 in the LV (c); bicaval view at balloon septostomy inflation (d).

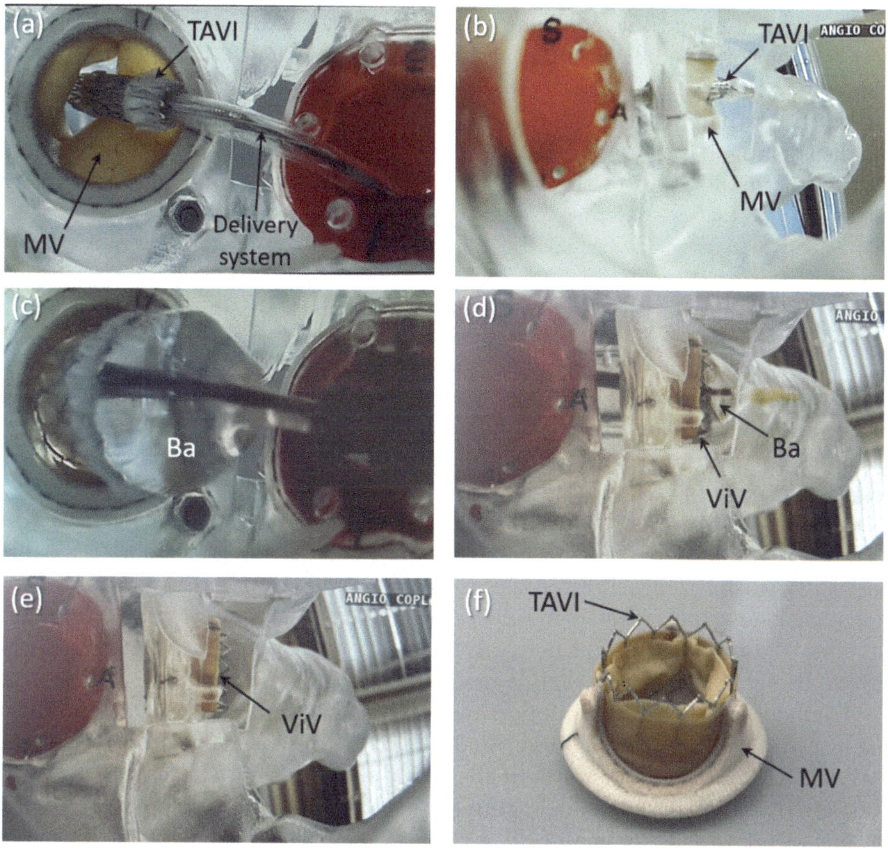

Fig. 5. Final valve in valve implantation: valves alignment from surgical (a) and angiographic coplanar (b); surgical (c) and angiographic coplanar (d) views at maximum balloon (Ba) inflation; details of the obtained valve in valve from angiographic coplanar view (e) and outside the model (f).

The design of cameras position and orientation was functional for the correct reproduction of the TEE and XA views (Fig. 2c), allowing the clinicians to navigate in the simulator and perform the procedural steps as for a real patient. In particular, during the TSP, the bicaval and the short axis views allowed the control of the transseptal introducer on the FO (Fig. 3a-b), checking the achieved position through the tenting and executing the TSP by the BRK (Fig. 3a-b). Concerning the surgical view, it was able to visualize the orientation of the Agilis catheter in the LA and its relative position with respect to the MV centre (Fig. 4a). In addition, by the angiographic coplanar view, it was possible to check the deployment of the pigtail catheter and the Safari[2] guide in the LV (Fig. 4b-c). The bicaval view allowed the axial positioning of the septostomy balloon through the FO and its inflation (Fig. 4d). Concerning the TAVI deployment, surgical and angiographic coplanar views allowed the axial and radial alignment between the collapsed TAVI and the surgical MV (Fig. 5a-b), the visualization of the balloon inflation (Fig. 5b-c) and the check of the valve in valve implantation (Fig. 5d). At the end of the simulation, it is possible to visualize the result by the disassembly of the MV (Fig. 5b-c). As regards the simulations, the operators completed the all steps of the procedure were correctly, without observing any issues.

4 Discussion

In this work a high-fidelity simulator for the reproduction of the entire TMViV procedure was designed, developed and tested by two expert clinicians. The fabrication of the anatomical components of the simulator was based on the elaboration of CT data and further CAD modifications to implement the real relative position of the structures and their rapid assembly/disassemble. A key feature is indeed the possibility of replacing the components independently, in case of damage and/or wear. In particular, the FO was designed as a disposable part to be easily replaced after the execution of TSP. Also the surgical mitral valve can be exchanged at convenience, in an easy and safe way. This allow to simulate the implantation of new valves on different types and sizes of surgical valves, by simply changing the LV model.

The adoption of cameras opportunely positioned and oriented to reproduce the intra-procedural imaging permitted the usage of the simulator outside of the cath lab without renouncing the realism of the real views. The inclusion of the apposite openings in the models were adequate to optimally visualize the same regions of interest as for real TEE and XA views, without interfering with the execution of the intervention.

Regarding the clinical aspect, the developed simulator allowed the exploitation of all the medical instrumentation involved during the TMViV procedure, from guidewires and catheters to delivery systems including proper TAVI valves. Actually, the usage of real valves may be interpreted as a limitation when these valves are unavailable. However, given the design of the system LA/LV for the housing of the valve (Fig. 1e), alternative solutions can be used, including valve fabrication with additive manufacturing or casting processes [1,3,10], thus avoiding the involvement of biological tissue.

While the preliminary tests conducted on the presented simulator demonstrated its feasibility to be considered as a promising tool for learning and training, few aspects need to be addressed. The anatomical components were fabricated as rigid, besides the FO part which instead was able to replicate a realistic haptic feedback during the TSP. Further improvements will include the adoption of deformable materials, especially for the RA model which interacts with the guidewires and transseptal introducer at the initial steps of the procedure. Candidate materials will be selected based on the market availability and considering the challenge to reproduce the mechanical properties of in-vivo tissue [4,6].

As further step, a campaign of tests would be beneficial to obtain feedback from a cohort of experts in TMViV. Future testing will increase the reliability of the simulator and its contribution to the increase of learning process, potentially resulting in improved clinical outcomes. Moreover, the design presented here could be easily translated for other types of transcatheter cardiovascular procedures. In conclusion, the development and testing of the presented patient-specific cardiovascular simulator mark a pivotal step in advancing interventional cardiology training. By bridging the gap between theoretical knowledge and practical skills, the simulator has the potential to significantly improve procedural success rates and patient outcomes in TMViV interventions.

Disclosure of Interests. The authors declare that they have no known competing financial interests or personal relationships that could have appeared to influence the work reported in this paper.

References

1. Antonuccio, M.N., et al.: Fabrication of deformable patient-specific AAA models by material casting techniques. Front. Cardiovasc. Med. **10**, 1141623 (2023)
2. Canavan, B., et al.: Transseptal mitral valve-in-valve-in-valve. CASE **8**(3), 253–257 (2024)
3. Cavallo, A., et al.: Fabrication and in-vitro characterization of a polymeric aortic valve for minimally invasive valve replacement. J. Mech. Behav. Biomed. Mater. **115**, 104294 (2021)
4. CeliS, S., et al.: 3D printing in modern cardiology. Curr. Pharm. Des. **27**(16), 1918–1930 (2021)
5. Ejiofor, J..I., et al.: Outcomes of repeat mitral valve replacement in patients with prior mitral surgery: a benchmark for transcatheter approaches. J. Thorac. Cardiovasc. Surg. **156**(2), 619–627 (2018)
6. Fanni, B.M., Antonuccio, M.N., Pizzuto, A., Berti, S., Santoro, G., Celi, S.: Uncertainty quantification in the in vivo image-based estimation of local elastic properties of vascular walls. J. Cardiovascu. Develop. Disease **10**(3), 109 (2023)
7. Harloff, M..T., et al.: A step-by-step guide to transseptal valve-in-valve transcatheter mitral valve replacement. Ann. Cardiothorac. Surg. **10**(1), 113 (2021)
8. Kikinis, R., Pieper, S.D., Vosburgh, K.G.: 3D Slicer: a platform for subject-specific image analysis, visualization, and clinical support. In: Intraoperative Imaging and Image-guided Therapy, pp. 277–289. Springer (2013)
9. Liu, Y., et al.: Transcatheter mitral valve-in-valve implantation applying a long pre-curved sheath for patients with degenerated bioprosthetic mitral valve. Rev. Cardiovasc. Med. **24**(2), 50 (2023)
10. Oveissi, F., Naficy, S., Lee, A., Winlaw, D., Dehghani, F.: Materials and manufacturing perspectives in engineering heart valves: a review. Mater. Today Bio **5**, 100038 (2020)
11. Quentin, V., et al.: Transcatheter mitral valve replacement using transcatheter aortic valve or dedicated devices: current evidence and future prospects. J. Clin. Med. **12**(21), 6712 (2023)

12. Schaefer, U., Frerker, C., Bader, R., Schmoeckel, M., Busse, C., Kuck, K.H.: Transcatheter access route options for treatment of degenerated mitral valve prosthesis with a balloon-expandable biological valve. Catheter. Cardiovasc. Interv. **82**(6), 999–1003 (2013)
13. Testa, L., Popolo Rubbio, A., Casenghi, M., Pero, G., Latib, A., Bedogni, F.: Transcatheter mitral valve replacement in the transcatheter aortic valve replacement era. J. Am. Heart Assoc. **8**(22), e013352 (2019)
14. Urena, M., Vahanian, A., Brochet, E., Ducrocq, G., Iung, B., Himbert, D.: Current indications for transcatheter mitral valve replacement using transcatheter aortic valves: valve-in-valve, valve-in-ring, and valve-in-mitral annulus calcification. Circulation **143**(2), 178–196 (2021)
15. Zafar, H., Soleimani, S., Ijaz, M., Zafar, J., Sharif, F.: Complex mitral valve anatomy and open issues in transcatheter mitral valve replacement. Surg. Practice Sci. **13**, 100182 (2023)

Design and Fabrication of 3D Printed Deformable Model of Patient-Specific Left Atrium for PIV Investigation

Emanuele Gasparotti[1(✉)], Benigno Marco Fanni[1], Emanuele Vignali[1], Mattia Dell'Amico[1,2], Rachele Marangoni[1,2], Francesca Danielli[3], Francesca Berti[3], Guido Parodi[4], Sergio Berti[5], Giancarlo Pennati[3], Lorenza Petrini[6], and Simona Celi[1]

[1] BioCardioLab, Bioengeneering Unit, Fondazione Toscana "G. Monasterio", Massa, Italy
{gasparotti,bmfanni,vignali,s.celi}@ftgm.it
[2] Department of Information Engineering, University of Pisa, Via Caruso, Pisa, Italy
[3] LaBS - Laboratory of Biological Structure Mechanics, Department of Chemistry, Materials and Chemical Engineering, Politecnico di Milano, Milan, Italy
[4] Cardiology Unit, Department of Medicine, Lavagna Hospital, Lavagna, Italy
[5] Adult Cardiology, Fondazione Toscana "G. Monasterio", Massa, Italy
[6] Department of Civil and Environmental Engineering, Politecnico di Milano, Milan, Italy

Abstract. Patients affected by non-valvular atrial fibrillation experience cardiac arrhythmia, which induces changes in blood flow and increases the risk of thrombus formation in the left atrial appendage (LAA), significantly raising the risk of stroke. Understanding the fluid dynamics within LAA is crucial for determining the most appropriate therapy for AF patients. In this study, an in vitro experimental setup was developed to study the fluid dynamics of the left atrium using a deformable phantom and a particle image velocimetry (PIV) system. A 3D-printed deformable phantom of the left atrium, including the LAA and pulmonary veins, was fabricated from a patient-specific computed tomography dataset. Additionally, a specific mock circulatory loop was established, with pulsatile flow imposed by a ventricular pump. PIV investigation was performed using a dedicated LED-PIV system, with the images analyzed focusing particularly on the LAA region.

Keywords: 3D printing · Left Atrial Appendage · PIV · Mock Circulatory Loop

1 Introduction

The left atrial appendage (LAA), is a cardiovascular structure, consisting in sac shape extrusion from the left atrium (LA). The LAA's unique anatomy [13] and its propensity to form blood clots have made it a focal point in studies related

© The Author(s), under exclusive license to Springer Nature Switzerland AG 2025
M. Carfagni et al. (Eds.): IDBN 2024, BioSysRob 35, pp. 81–94, 2025.
https://doi.org/10.1007/978-3-031-94626-4_10

to atrial fibrillation and stroke prevention [15]. The complex haemodynamics within the LAA can influence clot formation, making it crucial to understand these patterns in detail. Traditional imaging techniques are not able to capture with high detail the complex, three-dimensional flow patterns within the LAA. In this context, numerical and experimental simulations were able to retrieve a high-resolution vector maps of fluid motion, enabling detailed visualization and analysis of the flow [8–10]. In particular, as concerned experimental investigation, the mock circulatory loop embedded with anatomical phantom and particle image velocimetry (PIV) has emerged as a pivotal technique for the quantitative analysis of fluid dynamics within cardiac structures [7]. Concerning LAA investigation, the current state-of-art does not include PIV testing on complete and patient specific models of the LA and LAA. Given the difficulty of the physically realization of complex LA and LAA shapes, simpler solutions have been preferred. Some of these solutions involved the creation of models in which only the LAA is represented, which leads to unreliable results due to the oversimplification given by the absence of the LA contribution in the overall fluid dynamics. Others investigations presented PIV analyses on idealized LA phantom [6] or on patient-specific model realized in a material not matching completely the optical requirements of PIV [11].

This study aims to overcome such limitations and presents the design, fabrication, and testing of a three-dimensional, patient-specific model of the LA and LAA for use in PIV investigations.

2 Material and Methods

In this study, a new PIV simulator for the fluid dynamic investigation of the LA and LAA was designed, produced, and assessed. Starting from computed tomography (CT) images, a three-dimensional LA with LAA model was generated and then properly fabricated, by developing specific molds. The model was integrated into a mock circulatory loop, and hemodynamic characterization under physiological conditions was evaluated (Fig. 1). The study was conducted in accordance with the Declaration of Helsinki and approved by the review board of "Comitato Etico Regionale per la Sperimentazione Clinica della Toscana-Sezione Area Vasta Nord Ovest" (protocol code 23759_Berti).

2.1 Image Processing

The CT dataset from an adult patient who underwent a scan before LAAO procedure was used. The dataset consisted of 393 slices in the axial plane, with a resolution of 512 × 512 pixels and a slice thickness of 0.75 mm. A semi-automatic segmentation algorithm based on thresholding and manual adjustment was used to isolate the blood volume and to obtain the 3D geometry of the entire LA including LAA (Fig. 2).

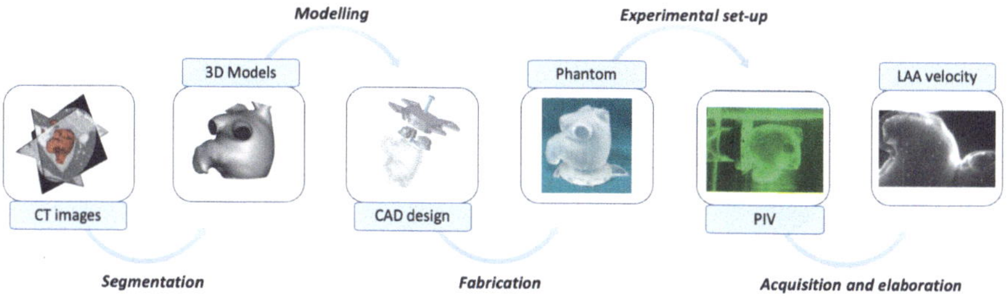

Fig. 1. Workflow

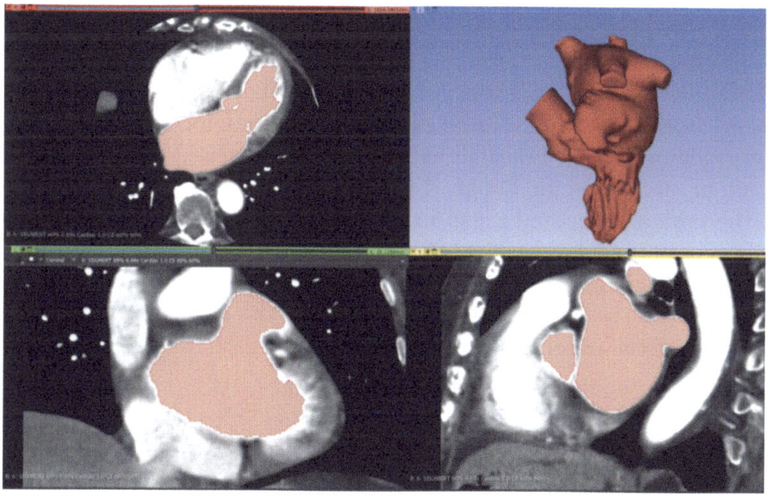

Fig. 2. Segmentation of the LH model (LV and LA) starting from CT scans.

2.2 Model Design

The CAD model of the phantom consisted in the LA and LAA surfaces of the segmented geometry, which were thickened outward by 3 mm (Fig. 3a). To connect the phantom to the mock circulatory loop, a cylindrical flange corresponding to the mitral annulus of the model was included. At the bottom of the flange, eight holes were designed with a depth of 5 mm and a diameter of 8 mm, intended to house cylindrical magnets for easy assembly and disassembly from the experimental setup. Given the phantom thickness and the dimensional requirements, a pouring technique based on inner and outer molds was adopted. The details of the design process are given in the following.

Outer Mold. The outer mold was designed from the outer surfaces of the phantom during the pouring procedure. Due to the presence of undercuts in the model, the outer mold was designed to allow an easier extraction of the

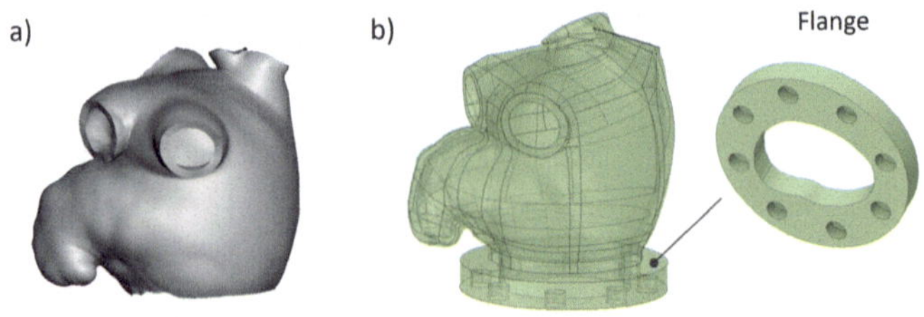

Fig. 3. LA phantom design: a) thickening of LA surfaces and b) LA model with flange

model and was manufactured in silicone material. A specific tank and a core mold (external core) were designed for the outer mold realization. The tank has a cylindrical shape and was splitted in two halves to facilitate the mold extraction. To reduce the volume of silicone needed for the outer mold two insert were defined (Fig. 4a). The external core comprised the outer surface of the phantom and a specific flange (Fig. 4b). The flange allowed positioning the core in the tank and created a gap in the outer mold. This gap allows phantom silicone to be poured in, forming the flange on the final model. An additional vertical pin was created at a pulmonary vein of the core to ensure the correct position between the core and the tank during the assembly (Fig. 4c). A 3D reconstrucion of the outer mold is shown in Fig. 4d.

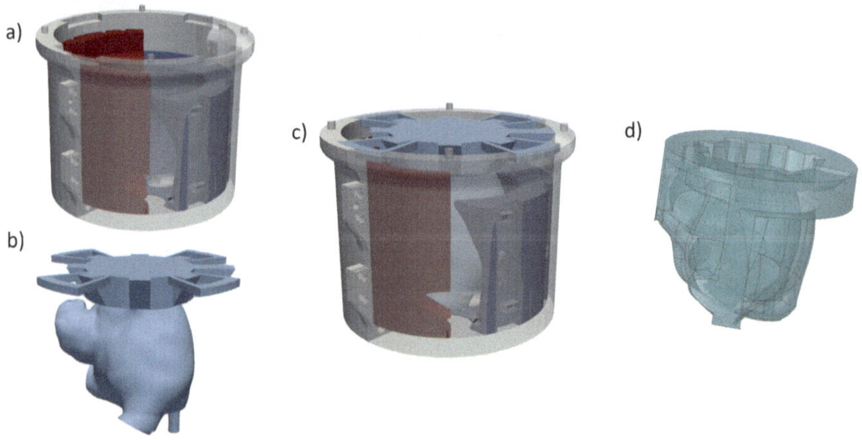

Fig. 4. Outer mold design: a) tank with outer inserts, b) external core, c) assembly for the realization of outer mold and d) outer mold

Inner Mold. The inner mold was designed starting from the inner surface of the phantom. In order guarantee the extraction of the inner mold from the phantom, an hollow, deformable and collapsible structure was designed. The inner mold was created from the inner surface of the phantom that was thickened inward of 5 mm. A collar and a flange were designed to hold the inner mold on the outer mold during the pouring phase. The collar is divided into two halves (Fig. 3a) to facilitate its removal from the phantom during the demolding process and is fastened to the flange by a bolt connection. The flange has eight cylindrical extrusions with a diameter of 8 mm and a height of 5 mm to create holes in the final phantom for the magnets (Fig. 3a). The assembly of the inner mold with the accessories is depicted in Fig. 3b.

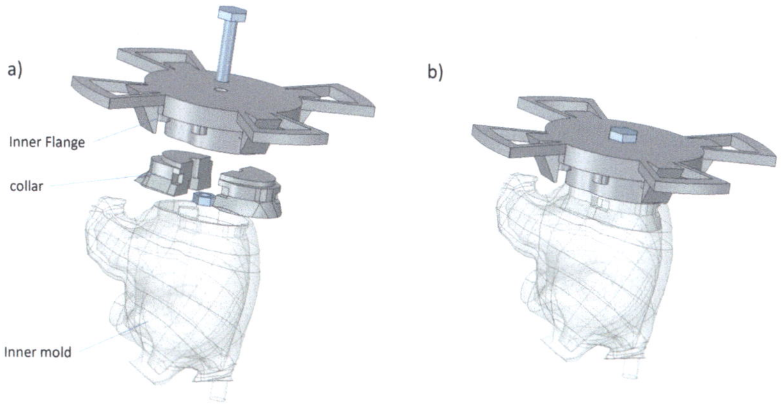

Fig. 5. Inner mold: a) Exploded view of the inner mold with collar and flange, b) Assembly of the inner mold for pouring phase

Due to the complexity of the inner mold, it was realized by designing two specific additional molds: the cavity core and the intermediate mold. The cavity core reproduces the internal wall of the inner mold, whereas the intermediate mold reproduces the external wall (Fig. 5). The overall scheme for its realization is depicted in Fig. 6.

Cavity Core - The cavity core was designed by connecting the internal wall of the inner mold to a reference flange, ensuring proper centering and providing access to the pouring channels for the inner mold (Fig. 6). Additionally, three rectangular cavities were created at 120 ° apart to allow the inner mold to be attached to the collar.

Intermediate Inner Mold Design - Intermediate inner mold is used to realize the external wall of the inner mold through a hollow structure. Due to the presence in the model of undercuts, the intermediate inner mold was designed to allow an easier extraction of the model and was manufactured in silicone material. For the

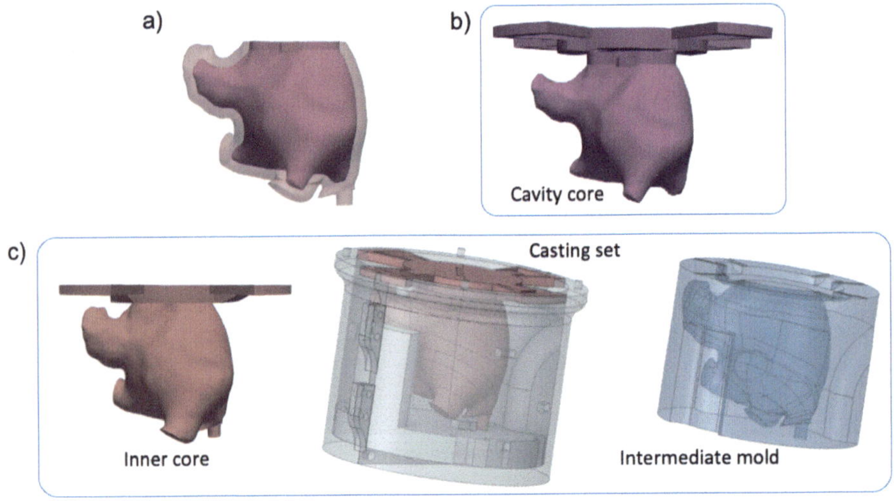

Fig. 6. Components for the realization of the inner mold: a) surfaces of inner mold, b) cavity core, c) inner core, tank, and inner inserts for the creation of intermediate mold

realization of the hollow in the mold, a inner core was designed, by connecting the external wall of the inner mold to a reference flange. The intermediate mold was realized by the assembly of the inner core with the tank, designed for the outer mold. To reduce the volume of silicone needed for the intermediate inner mold two specific inserts were defined.

2.3 Models Fabrication

Given the high complexity of the designed geometries, the molds, the accessories, and the model were realized combining 3D printing and silicone moulding techniques.

Outer Mold. The outer mold was manufactured using Mold Star 15 SLOW (Smooth-On, Macungie, USA) silicone. Moldstar is a soft silicone (shore 15A) characterized by high tear strength and high ultimate elongation. The accessories used for the outer mold were made in ABS material by using the Fused Deposition Modeling (FDM) technique with 3ntr A4v4 printer (3ntr, Oleggio, Italy). The printing parameters included a layer thickness of 0.1 mm and an infill of 40% for the external core, while a layer thickness of 0.2 mm was used for the tank and the inserts. All components were printed with an 25% of infill and an external perimeter composed by 4.5 outlines. The external core underwent to acetone vapors treatment for 1.5 h to reduce the surface roughness induced by the 3D printing process and make the surfaces suitable for PIV investigation. The external core, inserts, and tank were sprayed with two coats of release agent

(Ease Release 200, Smooth-On, Macungie, USA) to ensure the extraction of the outer mold.

Inner Mold. The inner mold was manufactured using Dragon Skin 10 Slow (Smooth-On, Macungie, USA) silicone, characterized by high ultimate elongation (1000%) and low shore hardness (10A). The intermediate mold was realized using Mold Star 15 SLOW. The accessories for the inner mold and intermediate mold were made from ABS material using the FDM technique. The printing parameters included a layer thickness of 0.1 mm and an infill of 40% for the cavity core, inner core, inner flange, and collar, while a layer thickness of 0.2 mm was used for the inserts. All components were printed with a 25% infill and an external perimeter composed of 4.5 outlines. The cavity core underwent acetone vapor smoothing. The cavity core, inner core, inserts, and tank were sprayed with two coats of Ease Release 200 to ensure the extraction of the intermediate and inner molds.

LA Phantom. The phantom was fabricated with optical silicone to ensure its compatibility with the PIV acquisitions. In particular, the Sylgard 184 (Dow, Wiesbaden, Germany), a two-component silicone material known for its optical properties [16], was used. The first phase involved coating the inner and outer molds with three layers of Ease Release 200 to ensure the phantom's extraction. The fabrication process consisted of several steps combining silicone mixing, degassing, and pouring. The phantom required 120g of material by mixing the silicone and the curing agent with a ratio of 10:1. The phantom required 120g of material, mixed at a 10:1 ratio of silicone to curing agent. After mixing, the casting material was degassed in a vacuum chamber at -0.8 bar for 30 min to remove air bubbles. The material was then poured into the molds in three steps to prevent air bubbles from being trapped in thick layers. After each pouring, a degassing phase at -0.8 bar was conducted. The casting phase was completed once the silicone reached the upper edge of the outer mold, ensuring no visible air bubbles remained. The silicone was cured for 48 h at room temperature.

2.4 PIV Setup

The fluid dynamic experiments on the phantom were performed using a mock circulatory loop and a PIV system. The setup overview is depicted in Fig. 7. The system includes: *i*) a reservoir where the LA phantom is positioned, *ii*) a versatile piston pump [2,12] to provide a pulsatile flow at mitral section of the phantom, *iii*) a flow sensor (CO.55/190 V2.0, Sonotec) to measure the mitral flow, *iv*) a CompactRIO for the system control and data acquisition and *v*) PIV system to capture to capture the fluid's velocity field within the phantom.

The PIV system consists by a pulsed high-power LED system (IL-106X LED Illuminator HardSOFT) used as the illumination source and a fiber optic line light (Schott A08589) with a cylindrical lens to form the light sheet. The images were acquired using a camera (acA1920-155μm - Basler Ace camera) with a 35 mm lens (RICOH FL-BC3518-9M). The camera and cylindrical lens are mounted on a truss structure that allows the adjustment of the relative positions of the camera and lens and sets the PIV system's height, depth, and position relative to the phantom.

A mixture of water (44%), gliceryne (34.5%) and urea (21.5%) was adopted to match the refractive index of Sylgard while simultaneously maintaining the density ρ and the dynamic viscosity μ of the blood [3]. The flow was seeded with 10 μm diameter hollow spherical particles to track the fluid velocity [18].

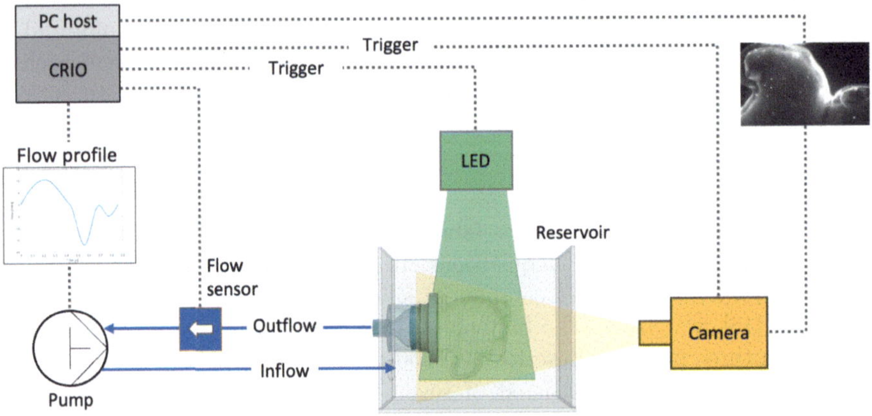

Fig. 7. Scheme of the mock circulatory loop with PIV system.

3 Results

Figure 8a-f show the results related to the realization of the outer mold (Fig. 8f), the outer core (Fig. 8a), the outer inserts (Fig. 8b), the tank (Fig. 8c). The FDM technique generated components without defects or distortions. All the complexities of the patient-specific geometry were reproduced. The external core and tank were easy to assemble (Fig. 8d-e). The outer core resulted successfully realized without any presence of defects related to incomplete curing or trapped air bubbles (Fig. 8d-e). It was necessary to cut the mold in two places to avoid damaging its surface during the extraction of the inner core. Figure 8g-i show the results related to the realization of the inner mold (Fig. 8l), the inner core (Fig. 8g), the inner inserts (Fig. 8h), the intermediate mold (Fig. 8i), the cavity core (Fig. 8j). The FDM technique was able to generate components without defects or dimensional distortions. In particular, all the complexities of the patient-specific geometry were reproduced. The inner core and tank were easy to assemble (Fig. 8h). The intermediate core resulted successfully realized without any presence of defects related to incomplete curing or trapped air bubbles

(Fig. 8i). It was necessary to cut the mold in two places to avoid damaging its surface during the extraction of the inner core. The cavity core and intermediate mold were easy to assemble (Fig. 8k) and the inner mold was realized without observable defects. Figure 9 depicts the inner mold (Fig. 9a), the molds assembly (Fig. 9b) and the resulting phantom (Fig. 9c). The inner and outer cores were easy to assemble, and the pouring process was facilitated by the casting channel in the outer molds. The LA phantom was realized without observable defects, outgrowths, or incomplete curing. The internal and external surfaces of the silicone model are smooth.

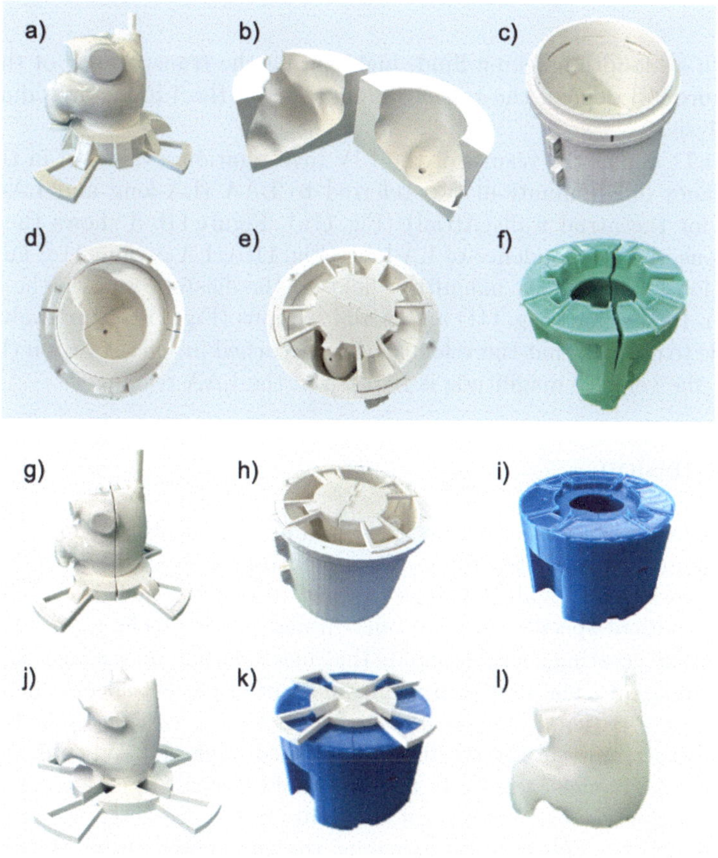

Fig. 8. Manufactured molds and accessories: a) outer core, b) outer inserts, c) tank, d) assembly of tank and outer inserts, e) assembly for the casting of outer mold, g) inner core, h) assembly of inner core, tank and inner inserts, i) intermediate mold, j) cavity core, k) assembly of cavity core and intermediate mold, l) inner mold

The manufactured model can be inserted into a MCL system to perform PIV investigations. Figure 10a shows the model placed within the dedicated reservoir

Fig. 9. LA phantom realization: a) assembly of inner mold, collar and flange, b) casting system, c) realized LA phantom

filled with a blood-mimicking fluid, highlighting the transparency of the phantom. Figure 10b depicts the model illuminated by the LED source during the PIV analysis.

Figure 11 depict the results of the PIV investigation, evaluated in three different planes of illumination: two referred to LAA (LA-long and LA-ostium) and one for the atrial wall (Atrial) (Fig. 11a). Figure 11b-d shows the camera acquisitions in corrispondence of LA-long (Fig. 11b), LA-ostium 11c) and Atrial 11d) sections. The velocity magnitude fields at the diastolic peak in the LA-long (Fig. 11e), LA-ostium (Fig. 11f) and Atrial sections (Fig. 11g) were evaluated. It is possible to observe that the velocity peak is reached in the LA region (Fig. 11e) and that the velocity magnitude is reduced in the LAA region.

4 Discussion

In the context of mock-circulatory loop systems, fabricating deformable models represents a crucial step for in-vitro quantitative analysis and device testing. This work presents all the steps required to manufacture deformable, PIV-compliant, patient-specific LA phantoms through casting techniques. These techniques involve creating reusable molds that meet specific dimensional and tolerance requirements. The first step in manufacturing patient-specific models was segmenting the patient's images. However, this step is user-dependent and can introduce uncertainty in the resulting 3D-printed model [4]. When dealing with patient-specific models, the CAD phase of mold design requires a certain level of expertise. The presence of several undercuts in the geometry necessitates the use of deformable molds to avoid damaging the final model during the extraction phase. Indeed, the design of molds requires the creation of various subcomponents. This design process includes evaluating the presence of undercuts, creating venting channels for air bubbles, establishing pouring channels, and defining the direction for casting extraction [1]. The FDM and silicone casting techniques allow for the production of components that meet dimensional and tolerance requirements. For the casting of the molds, a total of 1029 g of ABS material, 1450 g of Mold Star 15 Slow, and 90 g of Dragon Skin 10 Slow were used (Table 1). The realization of the outer and inner molds and their components requires

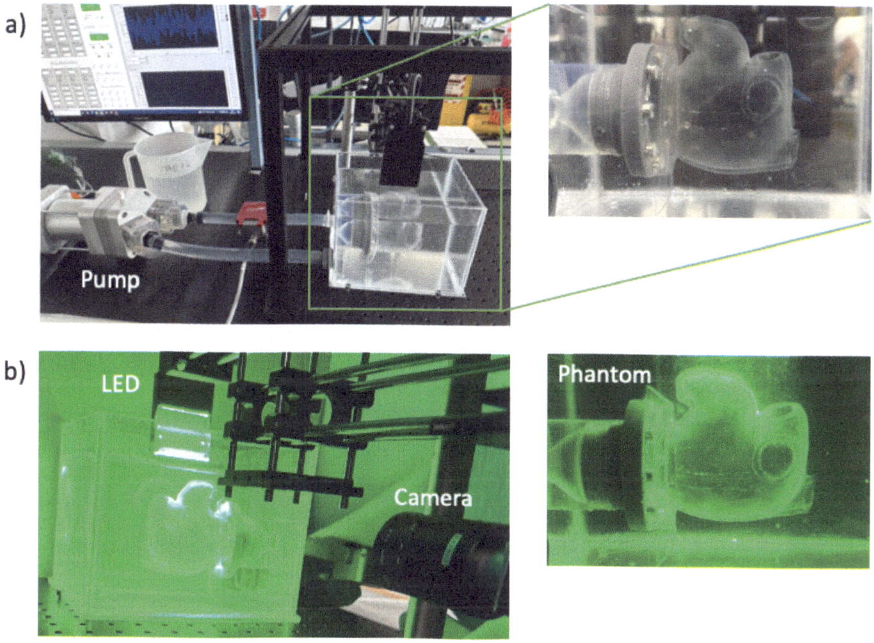

Fig. 10. PIV compatibility assessment: a) LA phantom connected to mock circulatory system and immersed in the working fluid, b) phantom behaviour under the LED sheet during PIV acquisition

92.5 h. The casting surfaces of the inner and outer molds were highly smooth, confirming that the FDM technique, combined with acetone smoothing, is an effective method for creating mold accessories. This casting strategy allows the molds to be reused to fabricate multiple copies of the same LA model.

Creating the phantom required a total of 76 h. The major drawback of Sylgard casting is its low viscosity and tendency to leak from the molds at the separation lines [1]. However, no leakages were observed when using silicone molds, and no sealants were required at the cuts of the outer mold. The phantom was easily removed from the outer core, and the hollow and soft nature of the inner core facilitated its easy extraction from the phantom. The phantom was fully cured, demonstrating that the silicones used for the molds, combined with the release agent, are compatible with Sylgard 184. No defects related to trapped air were observed, indicating that the degassing and casting steps were highly effective for producing a PIV-compatible phantom. The absence of bubbles prevents scatter artifacts in PIV applications. A preliminary hemodynamic assessment of the LA phantom was conducted by imposing physiological atrial fibrillation flow conditions through a custom mock circulatory loop and performing a velocimetry investigation using an LED-based PIV system. The phantom was transparent with the working fluid, allowing particle tracking in all selected sections. The velocity magnitude reached peaks of 0.45 m/s, similar to results from computa-

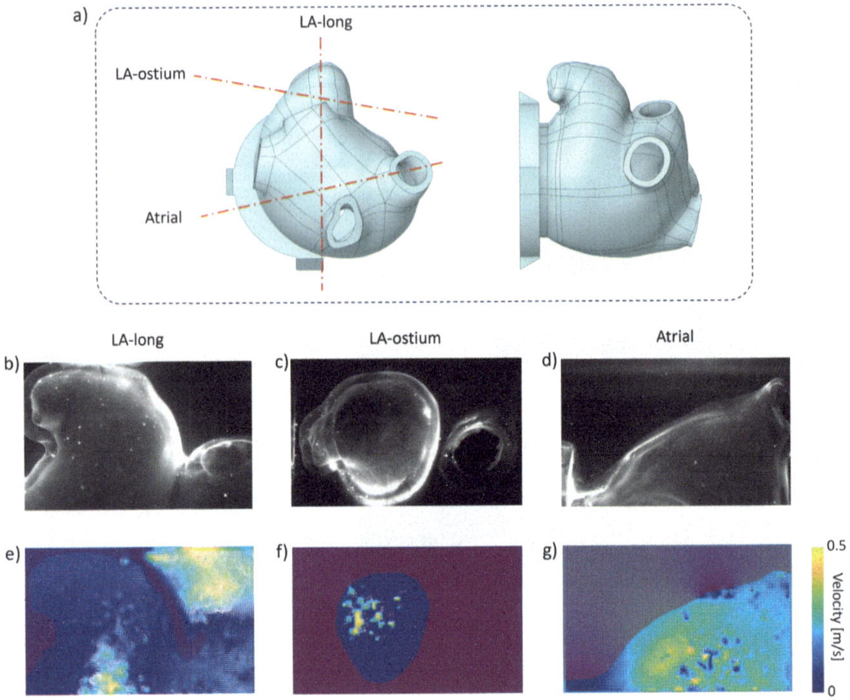

Fig. 11. Examples of PIV results in three different acquisition planes (a): LA-long (b,e); LA-ostium (c,f) and atrial (d,g).

Table 1. List of material used in the phantom realization.

Component	Material	Weight (g)	Fabrication time
Outer mold	Mold Star 15 Slow	600	4h 50 min
Intermediate mold	Mold Star 15 Slow	850	6h 45 min
Inner mold	Dragon Skin 10 Slow	90	7h 45 min
LA phantom	Sylgard184	120	76h
Tank	White ABS	369	35 h 50 min
Inserts (inner + outer)	White ABS	296	13 h 40 min
Outer core	White ABS	190	30 h 11 min
Inner core	White ABS	178	29 h 30 min
Cavity core	White ABS	112	13 h 40 min
Inner mold flange + collar	White ABS	70	5 h

tional fluid dynamic analyses in the literature [9,10,14]. The velocity decreased from the atrium to the LAA (Fig. 11e), as reported in the literature [8]. For the first time in the literature, a patient-specific and deformable 3D model with optical properties was created by combining 3D printing and silicone casting. The phantom was fully compatible with use in a mock circulatory loop for PIV investigation and refinement of numerical models. Future developments will focus on a comprehensive fluid dynamic investigation of the LAA at various depths. This will involve combining in-vitro experiments with in-silico simulations [5,17] to evaluate the changes in velocity associated with LAA treatment.

5 Conclusion

This study successfully demonstrated the design, fabrication, and testing of a patient-specific, three-dimensional model of the left atrium (LA) and left atrial appendage (LAA) for use in particle image velocimetry (PIV) investigations. Utilizing CT images for precise anatomical accuracy, the creation of the LA and LAA model involved advanced 3D printing and silicone casting techniques. The resultant model exhibited high fidelity to patient-specific geometries and optical properties suitable for PIV analysis. The integration of the model into a custom mock circulatory loop allowed for detailed hemodynamic assessment under physiological conditions. The PIV investigations revealed flow patterns and velocity magnitudes that are consistent with existing computational fluid dynamics (CFD) literature. This validation underscores the model's effectiveness in replicating realistic cardiac fluid dynamics, thus providing a valuable tool for refining numerical models and improving the understanding of LA and LAA hemodynamics.

Acknowledgments. This research was funded by the Italian Ministry of Health with the project PLACE (Planning LAa ClosurE), grant number RF-2021-12375208.

Disclosure of Interests. The authors declare that they have no known competing financial interests or personal relationships that could have appeared to influence the work reported in this paper.

References

1. Antonuccio, M.N., et al.: Fabrication of deformable patient-specific AAA models by material casting techniques. Front. Cardiovasc. Med. **10** (2023)
2. Bardi, F., Gasparotti, E., Vignali, E., Avril, S., Celi, S.: A hybrid mock circulatory loop for fluid dynamic characterization of 3D anatomical phantoms. IEEE Trans. Biomed. Eng. **70**(5), 1651–1661 (2022)
3. Brindise, M.C., Busse, M.M., Vlachos, P.P.: Density-and viscosity-matched newtonian and non-newtonian blood-analog solutions with PDMS refractive index. Exp. Fluids **59**, 1–8 (2018)
4. Celi, S., Vignali, E., Capellini, K., Gasparotti, E.: On the role and effects of uncertainties in cardiovascular in silico analyses. Front. Med. Technol. **3**, 748908 (2021)

5. Danielli, F., et al.: Left atrial appendage occlusion: on the need of a numerical model to simulate the implant procedure. Int. J. Num. Meth. Biomed. Eng. **40**(5), e3814 (2024)
6. Dueñas-Pamplona, J., Sierra-Pallares, J., García, J., Castro, F., Munoz-Paniagua, J.: Boundary-condition analysis of an idealized left atrium model. Ann. Biomed. Eng. **49**(6), 1507–1520 (2021)
7. Falahatpisheh, A., Kheradvar, A.: High-speed particle image velocimetry to assess cardiac fluid dynamics in vitro: from performance to validation. Europ. J. Mech. B/Fluids **35**, 2–8 (2012). cardiovascular Flows
8. Fanni, B.M., et al.: Correlation between LAA morphological features and computational fluid dynamics analysis for non-valvular atrial fibrillation patients. Appl. Sci. **10**(4), 1448 (2020)
9. Gasparotti, E., et al.: Computational fluid dynamic simulation to evaluate the device-related effects after left atrial appendage occlusion. In: International Symposium on Computer Methods in Biomechanics and Biomedical Engineering, pp. 205–212. Springer (2023)
10. Mill, J., et al.: Sensitivity analysis of in silico fluid simulations to predict thrombus formation after left atrial appendage occlusion. Mathematics **9**(18), 2304 (2021)
11. Soroosh, S.: Hemodynamic Indices and Shape-Based Models of Left Atrial Appendage to Enhance Stroke Prediction in Atrial Fibrillation. Ph.D. Thesis, University of Pittsburgh (2021)
12. Vignali, E., Gasparotti, E., Mariotti, A., Haxhiademi, D., Ait-Ali, L., Celi, S.: High-versatility left ventricle pump and aortic mock circulatory loop development for patient-specific hemodynamic in vitro analysis. ASAIO J. **68**(10), 1272–1281 (2022)
13. Vivoli, G., et al.: Simultaneous functional and morphological assessment of left atrial appendage by 3D virtual models. J. Healthcare Eng. **2019**(1), 7095845 (2019)
14. Wang, L., Wang, Z., Fang, R., Li, Z.Y.: Evaluation of stroke risk in patients with atrial fibrillation using morphological and hemodynamic characteristics. Front. Cardiovasc. Med. **9**, 842364 (2022)
15. Yaghi, S., et al.: Left atrial appendage morphology improves prediction of stagnant flow and stroke risk in atrial fibrillation. Circul . Arrhyth. Electrophys. **13**(2), e008074 (2020)
16. Yousif, M.Y., Holdsworth, D.W., Poepping, T.L.: Deriving a blood-mimicking fluid for particle image velocimetry in sylgard-184 vascular models. In: 2009 Annual International Conference of the IEEE Engineering in Medicine and Biology Society, pp. 1412–1415. IEEE (2009)
17. Zaccaria, A., et al.: Left atrial appendage occlusion device: development and validation of a finite element model. Med. Eng. Phys. **82**, 104–118 (2020)
18. Zhang, R., Zhang, Y.: An experimental study of pulsatile flow in a compliant aortic root model under varied cardiac outputs. Fluids **3**(4) (2018)

CAD-Driven Design for 3D-Printed Surgical Guides and Cranial Implants

Francesco Buonamici[1(✉)], Monica Carfagni[1], Federico Mussa[2], Agushi Rina[2], and Antonio Marzola[3]

[1] Department of Industrial Engineering of Florence, Via di S. Marta 3, 50139 Florence, Italy
francesco.buonamici@unifi.it

[2] Meyer Children's Hospital IRCCS, Viale Pieraccini 24, 50141 Florence, Italy

[3] Department of Industrial and Information Engineering and Economics, University of L'Aquila, 67040 L'Aquila, Italy

Abstract. This paper presents a novel workflow for designing and fabricating patient-specific cranial implants and surgical guides using advanced CAD and 3D printing technologies. Developed at the T3Ddy Lab at Meyer Children's Hospital, the methodology includes close collaboration between engineers and neurosurgeons, allowing for rapid prototyping and real-time adjustments. The workflow involves designing a cutting guide for precise bone resection and a custom cranial implant for anatomical reconstruction, both using titanium alloy for its biocompatibility and mechanical strength. The setup enhances efficiency, ensuring precise device fit and improving surgical outcomes. The paper presents the general framework established to design custom made medical devices for skull repairing surgery. The CAD modeling phase of the surgical guide and cranial implant are described in detail, highlighting developed solutions to improve the effectiveness of the technology. The proposed workflow was validated in multiple clinical cases, demonstrating improved surgical precision, reduced operative time, and enhanced recovery.

Keywords: Craniotomy · Cranioplasty · 3D Printing · Surgical Guide · CAD · Cranial Implant

1 Introduction

Craniotomy and cranioplasty are two crucial neurosurgical procedures involving the manipulation of the skull to access or reconstruct the brain and its surrounding structures. These operations are often required after trauma, tumors, or other medical conditions, and involve either removing a portion of the skull to access the brain (craniotomy) or reconstructing the skull (cranioplasty) to restore its protective function and aesthetic appearance.

In recent years, there has been growing interest in leveraging advanced technologies such as Computer-Aided Design (CAD) and 3D printing to improve the outcome of these procedures [1]. This article explores the process of designing patient-specific

© The Author(s), under exclusive license to Springer Nature Switzerland AG 2025
M. Carfagni et al. (Eds.): IDBN 2024, BioSysRob 35, pp. 95–101, 2025.
https://doi.org/10.1007/978-3-031-94626-4_11

cranial implants and surgical guides using a CAD driven approach that was developed at T3Ddy lab, the 3D printing lab of the Meyer's Children Hospital of Florence. The methodology described in the present work encompasses novel key aspects contributing to the effectiveness of the devices produced and the overall well-being of the patient.

The design of personalized surgical tools for cranioplasties has been studied for a long time [2, 3], with numerous contributions and published works that cover various aspects of the field, including advancements in materials, methods for improving surgical accuracy, and approaches for reducing patient recovery times. Among the most recent advancements in this area, the use of low-cost and accessible 3D printing technologies and the algorithm-driven optimization of the implants to satisfy specific goals are among the emerging research trends. According to Czyżewski et al. (2022), polymer-based endoprostheses have proven to be reliable and significantly cheaper than conventional titanium implants, making them accessible for hospitals in low-income countries. The use of materials such as PEEK for the fabrication of cranial implants has opened new possibilities for custom solutions in neurosurgery. As previously mentioned, another area that has observed a lot of research interest is the development of optimization techniques for the design of cranial implants. Multi-objective optimization algorithms or AI-driven tools have been proposed to guide designers towards the maximization of mechanical and structural properties of cranial implants, often improving both functional and aesthetic outcomes. By incorporating complex geometries such as Triply Periodic Minimal Surfaces (TPMS), these implants can be designed to achieve enhanced mechanical strength and better integration with surrounding tissue, ultimately leading to more successful surgical outcomes. For the interested reader, a recent review on the application of CAD techniques for cranioplasty treatments is presented in (Thimukonda Jegadeesan et al., 2022).

Less attention has been given to understanding the impact that different design features of custom-made implants have on their usability as perceived by surgeons during operations. Ergonomic features and geometric characteristics of patient-specific devices used in reconstructive cranioplasties must be optimized to enhance their effectiveness, ease of use, and ultimately to minimize the risk of errors in the operating room. This paper focuses on such research question, presenting a comprehensive methodology for fabricating patient-specific titanium cranial plates and cutting guides for cranioplasties. The proposed method was validated through several clinical cases at Meyer Children's Hospital, demonstrating its reliability and practical value.

2 Methodology

The methodology hereafter described aims at the design of two patient-specific devices to be used in a reconstructive craniotomy: a cutting guide, identifying the portion of the skull to be cut, and a cranial implant, to be used for the final reconstruction of the anatomy. The workflow, depicted in Fig. 1, combines both physical and digital techniques to achieve the best possible outcome, ensuring that the devices meet both functional and aesthetic requirements. The figure highlights the surgeon's contribution in the planning steps. A close collaboration between engineers and clinicians is fundamental to reduce design changes to be implemented in later stages of production.

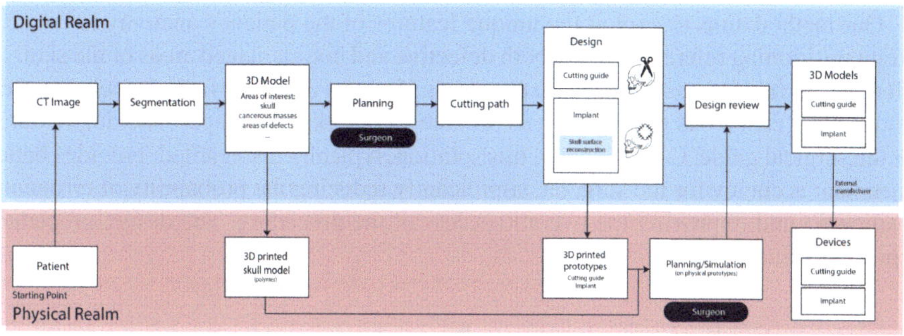

Fig. 1. Workflow of the approach developed at T3Ddy laboratory.

While the overall approach may be considered standard, as it has been developed and adopted by multiple clinical centers over the last decade, the proposed procedure distinguishes itself in several key areas:

- The use of a titanium alloy for both the implant and the cutting guide, enabling specific features of the final devices [4]. This solution is typically disregarded due to higher costs when developing patient-specific devices for craniotomies and especially surgical guides, but it allows for the fabrication of really thin features while controlling deflections.
- The design solutions used for the surgical guide, discussed later in the text, which have improved the effectiveness of the tool for the surgeon.
- The proximity of T3Ddy Lab to the surgical teams allows for real-time adjustments, rapid prototyping, and iterative improvements to the surgical tools and implants, ultimately resulting in improved surgical times. The design features of the devices are then transferred to an external manufacturing service, which receives them as part of the medical prescription that the surgeon produces, according to 2017/745 regulation (EU).

The procedure will be discussed in the following sections in detail, focusing on the design features and CAD solutions used. Regarding the software tools used throughout the procedure, segmentation and planning are carried out using Materialise Mimics; modeling steps are mainly performed within Geomagic Design X, a CAD modeling software capable of working with polygonal models; additional operations are performed within NX, a parametric associative CAD modeler developed and commercialized by Siemens.

The devices are manufactured in Ti6Al4V using Electron Beam Melting technology by external producers in compliance with UE 2017/745.

2.1 Cutting Guide

The design of the cutting guide takes as input the anatomy of the skull, reconstructed during the segmentation phase, and the cutting path produced by the surgeon. A significant number of cranioplasty procedures are performed to pursue the removal of cranial bone tumors (Czyżewski et al., 2022), which produce notable deformations.

Our method aims to exploit the unique features of the patient's anatomy to create a strong positioning reference, using both defective and non-damaged areas of the skull as reference surfaces to ensure accurate matching. High curvature surfaces, which are often produced by cancerous tissue, serve as reliable landmarks and shape-matching features for the surgical guide. Consequently, this solution, typically disregarded, provides better placement accuracy for the surgeon, significantly reducing the probability of erroneous positioning and improving the overall success of the procedure. The design procedure is presented in Fig. 2.

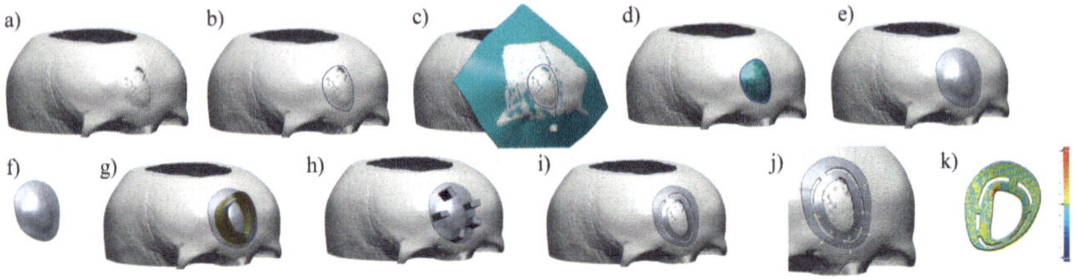

Fig. 2. Design workflow for the surgical guide. a) segmented anatomy – STL model, clear deformation produced by the tumor; b) cutting path requested by the surgeon; c) CAD surface fitted to the anatomy; d) surface patch obtained by cutting c) with d); e) and f) surgical guide main volume; g) series of extruded and offset surfaces used to generate the cutting path and internal negative volume; h) CAD bodies modeled to introduce connecting tabs on the guide; i) final 3D model of the guide; j) and k) quality control.

The surgical guide is designed from a continuous CAD surface fitted on the anatomical data (segmented STL model) of the skull in the region of interest (Fig. 2c). This surface is cut with multiple curves derived from the cutting path proposed by the surgeon (Fig. 2b) and extruded to produce a constant-thickness body, which forms the base of the surgical guide. Thicknesses ranging from 0.6 to 2 mm can be used, depending on the desired stiffness. The reduced thickness, allowed by the use of a Titanium metal alloy as manufacturing material, reduces the risk of interference between the surgical tools and the cutting template. The surgical guide slot is 2 mm wide, a dimension that allows the use of a surgical marker to trace the cutting path directly on the bone. Alternatively, the cranial drill can be guided into the slot. For safety reasons, the slot is not centered directly on the cutting path; instead, the cutting path corresponds to the inner profile of the slot. This design choice minimizes the risk of erroneous interpretation of the guide's indications, ensuring safer surgical procedures.

Lateral profiles of the surfaces of the surgical guides are generated offsetting the extruded cutting path surface. Connecting tabs between are added by manually modeling the required bodies (Fig. 2h) to bridge the cutting slot region.

Quality control checks are applied at the end of the procedure to validate the design. Specifically, possible interferences measured between the device and the patient's bone should be kept under 0.1 mm. Local deviations can be accepted as long as they do not produce interference; they are occasionally encouraged to avoid contact in delicate areas: where the diagnostic image is not clear, where the anatomy of the bone is too complex and fragmented and a clean contact could not have been guaranteed.

2.2 Cranial Plate

The cranial plate is designed following a similar workflow, presented in Fig. 3, but requires additional steps to reconstruct the missing portion of the skull. For cases involving previous craniotomies, such as after the removal of cancerous bone tissue or trauma-induced defects, a crucial step is the reconstruction of the shape of the skull to restore its form and function. Different strategies have been proposed to achieve this [5]. Whenever possible, contralateral and symmetrical portions of the skull can be used to constraint the fitting procedure, allowing for a natural fit of the implant. Additionally, previous medical images may provide valuable data. When direct anatomical references are unavailable, advanced tools such as Statistical Shape Models (SSM) [6] can be employed to generate a plausible physiological shape based on statistical anatomical data. Also, geometrical tools analyzing the curvature of the available part of the skull can be integrated in the process.

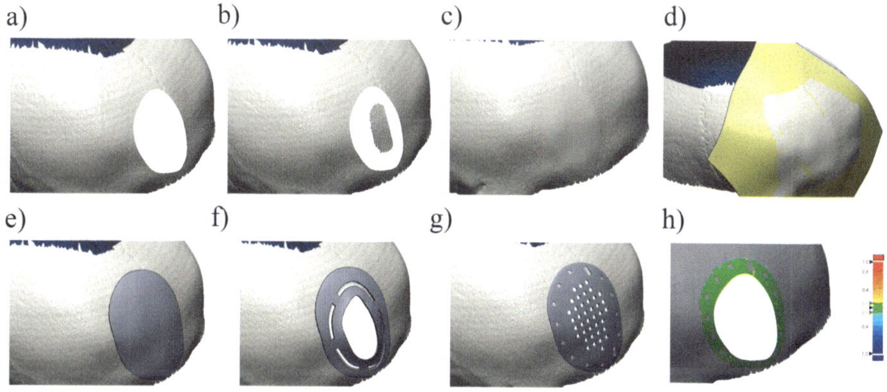

Fig. 3. Design workflow for the cranial implant; a) starting polygonal surface with the cut region removed; b) contralateral surface mirrored in the defect area to guide the fitting; c) reconstructed polygonal skull surface; d) reconstructed CAD surface of the skull; e) main body of the cranial implant; f) corresponding surgical guide (identical external profile) aligned with the removed portion of the skull; g) final CAD model of the cranial plate with advanced design features; h) quality control with a 0.1 mm tolerance zone (green).

Throughout the years, T3Ddy has implemented a range of these strategies (e.g. [6, 7]), adapting the approach to the patient's specific needs. The reconstruction often involves a combination of semi-automatic CAD techniques and manual adjustments by experienced users.

The cranial plate's external profile is designed to match the external contour of the surgical cutting guide, aiding in the precise positioning of the implant. During the first part of the intervention, the surgeon usually marks also the external profile of the guide, which can be used later to position the implant. This assures an accurate placement even without the guidance of the defective region – no longer present after resection. The final modelling of the cranial plate involves thickening the reconstructed reference surface

(Fig. 3d) and trimming it using a tool surface obtained from the extrusion of the plate's external profile (Fig. 3e).

The main body of the implant is then refined introducing additional features:

- hollowing holes, in a regular pattern covering the internal part of the plate, for lightweighting purposes;
- a set of holes on the outside profile of the plate, distanced 8 to 12 mm from the external profile of the plate, for primary fixation of the device with surgical screws;
- optional reference features for positioning (e.g. slots in Fig. 3g) for alignment with cranial suture line)

3 Discussion and Conclusions

This paper has presented a comprehensive workflow for designing and fabricating patient-specific cranial implants and surgical guides using advanced CAD and 3D printing technologies. Developed at the T3Ddy Lab at Meyer Children's Hospital, this approach emphasizes close collaboration between surgeons and engineers, allowing for the creation of custom surgical tools that are both functional and practical for clinical use.

The use of titanium alloy allows for the fabrication of thin structures that would otherwise be unattainable given the required stiffness of the device. For cranial implants, this represents a significant advantage over polymer-based devices, enabling a mode of use in which the device is applied directly onto the cranial vault, supported by the lateral edges of the defect. This approach ensures stability and proper alignment while minimizing the risk of shifting during surgery. In other cases, devices may be placed directly within the skull opening, but this requires a very precise fit to ensure that, once the procedure is completed, there are no visible imperfections or gaps left between the implant and the surrounding bone. Any misalignment or insufficient coverage can lead to aesthetic concerns and may also compromise the mechanical stability of the implant. Ensuring a perfect fit is therefore crucial to achieving both a cosmetically acceptable outcome and a structurally sound reconstruction, with effective integration of the implant into the existing cranial anatomy. The presented procedure offers a more forgiving approach to the surgeon, which can benefit from an easier cutting operation.

The method was tested on 5 cases at the Meyer Children Hospital of Florence; the application of the presented workflow has demonstrated significant improvements in surgical outcomes, including reduced operative time, enhanced accuracy of bone resection, and better-fitting cranial implants. By leveraging modern CAD software and in-hospital 3D printing, the workflow offers an efficient and accessible solution for cranial reconstruction, particularly in resource-constrained environments.

Future research should focus on optimizing the design workflow to minimize the manual workload involved. Current processes still require significant manual intervention to integrate different modeling tools and ensure quality control. Streamlining these steps through automation and improving software interoperability could enhance efficiency and reduce the potential for human error, making personalized cranial implant production more practical and scalable. AI tools could play a crucial role in automating complex modeling tasks [8], enhancing design optimization, and providing predictive insights to further refine the production process. These advancements could make the entire workflow more efficient, allowing for broader adoption and better accessibility of personalized implants in diverse clinical settings.

References

1. Thimukonda Jegadeesan, J., Baldia, M., Basu, B.: Next-generation personalized cranioplasty treatment. Acta Biomater. **154**, 63–82 (2022). https://doi.org/10.1016/j.actbio.2022.10.030
2. Maravelakis, E., David, K., Antoniadis, A., Manios, A., Bilalis, N., Papaharilaou, Y.: Reverse engineering techniques for cranioplasty: a case study. J. Med. Eng. Technol. **32**(2), 115–121 (2008). https://doi.org/10.1080/03091900600700749
3. Jardini, A.L., Larosa, M.A., de Carvalho Zavaglia, C.A., Bernardes, L.F., Lambert, C.S., Kharmandayan, P., Calderoni, D., Maciel Filho, R.: Customised titanium implant fabricated in additive manufacturing for craniomaxillofacial surgery: This paper discusses the design and fabrication of a metallic implant for the reconstruction of a large cranial defect. Virtual Phys. Prototyp. **9**(2), 115–125 (2014). https://doi.org/10.1080/17452759.2014.900857
4. Luo, J., Morrison, D.A., Hayes, A.J., Bala, A., Watts, G.: Single-piece titanium plate cranioplasty reconstruction of complex defects. J. Craniofac. Surg. **29**(4), 839–842 (2018). https://doi.org/10.1097/SCS.0000000000004311
5. Buonamici, F., Furferi, R., Genitori, L., Governi, L., Marzola, A., Mussa, F., Volpe, Y.: Reverse engineering techniques for virtual reconstruction of defective skulls: An overview of existing approaches. Comput-Aided Des. Appl. **16**(1), 103–112 (2018). https://doi.org/10.14733/cadaps.2019.103-112
6. Marzola, A., Servi, M., Volpe, Y.: A Reliable Procedure for the Construction of a Statistical Shape Model of the Cranial Vault, pp. 788–800. Springer, Cham (2019). https://doi.org/10.1007/978-3-030-31154-4_67
7. Marzola A, McGreevy KS, Mussa F, Volpe Y, Governi L (107516) HyM3D: A hybrid method for the automatic 3D reconstruction of a defective cranial vault.Comput. Methods Programs Biomed. Volume 234 234:. doi:https://doi.org/10.1016/j.cmpb.2023.107516
8. Kodym, O., Li, J., Pepe, A., Gsaxner, C., Chilamkurthy, S., Egger, J., Španěl, M.: SkullBreak / SkullFix—Dataset for automatic cranial implant design and a benchmark for volumetric shape learning tasks. Data Brief. **35**, 106902. https://doi.org/10.1016/j.dib.2021.106902

Author Index

A
Asero, Riccardo 61

B
Baiamonte, Giuliana 61
Berti, Francesca 81
Berti, Sergio 69, 81
Bertolini, Michele 25
Bolli, Eleonora 45
Buonamici, Francesco 95

C
Calì, Michele 61
Capelli, Claudio 25
Carfagni, Monica 17, 95
Carpi, Federico 9, 33
Celi, Simona 69, 81
Colombo, Giorgio 25
Corno, Chiara 69
Credi, Caterina 9, 33
Cvetkovic, Mirjana 25

D
Danielli, Francesca 81
De Gaetano, Francesco 25
De Luca, Marco 17
Deidda, Virginia 9
Dell'Amico, Mattia 81
Di Martino, Gianfranco 61

F
Fanni, Benigno Marco 69, 81
Fava, Alessandra 53
Feregotto, Greta 33

G
Gasparotti, Emanuele 69, 81

K
Kaciulis, Saulius 45, 53

L
Langione, Marianna 9
Lanzutti, Alex 45
Laudani, Giuseppe 61

M
Marangoni, Rachele 81
Marchini, Alessandro Maria 25
Mariani, Massimiliano 69
Martella, Daniele 33
Marzola, Antonio 17, 95
Mencarelli, Marta 17
Mezzi, Alessio 45, 53
Monaco, Fulvia 61
Montanari, Roberto 45, 53
Murthy, Neel 25
Mussa, Federico 95

P
Pallotta, Stefania 1
Palombi, Alessandra 45, 53
Parmeggiani, Camilla 33
Parodi, Guido 81
Pennati, Giancarlo 81
Petrini, Lorenza 81
Pioner, José Manuel 9
Profili, Andrea 1

R
Rina, Agushi 95
Rossi, Ruggero 33

S
Servi, Michaela 1
Simonetti, Giovanni 33
Smith, Luke 25
Sordetti, Francesco 45

T
Torzini, Lorenzo 17

© The Editor(s) (if applicable) and The Author(s), under exclusive license to Springer Nature Switzerland AG 2025

M. Carfagni et al. (Eds.): IDBN 2024, BioSysRob 35, pp. 103–104, 2025.
https://doi.org/10.1007/978-3-031-94626-4

V
Vaglio, Emanuele 45
Varone, Alessandra 45, 53
Ventisentte, Isabel 9
Verona, Claudio 45
Vignali, Emanuele 81
Volpe, Yary 1, 17

Z
Zahid, Ali 45
Zola, Roberta 69

The manufacturer's authorised representative in the EU is Springer Nature Customer Service Centre GmbH, Europaplatz 3, 69115 Heidelberg, Germany. If you have any concerns regarding our products, please contact ProductSafety@springernature.com

Printed and bound by CPI Group (UK) Ltd, Croydon, CR0 4YY
01/10/2026
02225944-0002